THE STEAM PLANET

Life and Atmosphere on K2-18b

HARIKUMAR V T

PREFACE

The discovery of exoplanets has transformed our understanding of the universe and our place within it. Among these distant worlds, K2-18b stands out as a beacon of hope and intrigue. Located 124 light-years away, this super-Earth has captured the imaginations of scientists and the public alike with its steamy atmosphere and the tantalizing possibility of water, the essential ingredient for life as we know it.

"The Steam Planet: Life and Atmosphere on K2-18b" delves into the fascinating world of this extraordinary exoplanet. This book is a journey through the discovery, exploration, and ongoing study of K2-18b, a planet that challenges our understanding of habitability and the potential for life beyond Earth.

In these pages, we will explore the origins of K2-18b's discovery by NASA's Kepler spacecraft and the subsequent observations that have revealed a rich tapestry of atmospheric phenomena. We will examine the groundbreaking detection of water vapor in its atmosphere, a finding that has sparked both excitement and debate within the scientific community.

Through the lens of K2-18b, this book will also illuminate the broader context of exoplanet research. We will discuss the advanced technologies and methods that have made such discoveries possible and the future missions poised to further unravel the mysteries of distant worlds. Importantly, we will ponder the implications of these discoveries for the age-old question: Are we alone in the universe?

"The Steam Planet" is not just a scientific exploration but also a celebration of human curiosity and the relentless pursuit of knowledge. It highlights the collaborative efforts of astronomers, astrophysicists, and space agencies around the world, all working together to push the boundaries of what we know.

Whether you are a seasoned astronomer, a science enthusiast, or simply a curious reader, this book aims to provide a comprehensive and accessible account of one of the most exciting frontiers in modern science. As we journey together through the wonders of K2-18b, may we be reminded of the vastness of the cosmos and the endless possibilities that await us.

Welcome to the world of K2-18b, a steamy planet that may hold the keys to unlocking the secrets of life beyond our solar system.

COPYRIGHT WARNING

Copyright © 2024 HARIKUMAR V T All rights reserved.

No part of this publication may be reproduced, distributed, or transmitted in any form or by any means, including photocopying, recording, or other electronic or mechanical methods, without the prior written permission of the publisher, except in the case of brief quotations embodied in critical reviews and certain other noncommercial uses permitted by copyright law.

The author and publisher disclaim all responsibility for any liability, loss, or risk, personal or otherwise, which may be incurred as a consequence, directly or indirectly, of the use and application of any of the contents of this book.

For permissions requests, inquiries about licensing, and other copyright-related matters, please contact:

HARIKUMAR V T

[vtharipnra@gmail.com]

Thank you for respecting the hard work and intellectual property rights of the author.

CONTENTS

1. Discovery of K2-18b: The groundbreaking moment when astronomers first identified K2-18b among the stars.

2. The Science of Exoplanets: Understanding the methods and technologies that enable us to discover and study distant worlds.

3. A Red Dwarf's Companion: Exploring K2-18b's host star, K2-18, and its influence on the planet.

4. Super-Earths: A New Class of Planets: Defining super-Earths and what makes K2-18b unique among them.

5. Atmospheric Revelations: How scientists detected water vapor in K2-18b's atmosphere and what it means.

6. The Steamy Skies of K2-18b: Diving into the composition and characteristics of the planet's steamy atmosphere.

7. Climate and Weather on K2-18b: Speculating on the dynamic weather patterns and climate of a steam-filled world.

8. Habitability: Beyond the Goldilocks: Zone Examining the factors that could make K2-18b a candidate for habitability.

9. Water Worlds: A New Frontier: Comparing K2-18b to other known and theorized water worlds in the universe.

10. Technological Marvels: Tools of Discovery: Highlighting the telescopes and instruments that have unveiled the secrets of K2-18b.

11. The Role of the James Webb Space Telescope: How the JWST will advance our understanding of K2-18b and similar exoplanets.

12. Life as We Know It Possibilities on K2-18b: Speculating on the potential forms of life that could exist in a steamy atmosphere.

13. Future Missions and Observations: The upcoming space missions and technologies set to explore K2-18b further.

14. K2-18b in Popular Culture: How the discovery of K2-18b has captured the imagination of the public and inspired media.

15. The Broader Implications: Are We Alone? : Reflecting on what the study of K2-18b means for the search for extraterrestrial life and our place in the cosmos.

1. DISCOVERY OF K2-18B

The groundbreaking moment when astronomers first identified K2-18b among the stars.

The universe, with its infinite expanse and myriad celestial bodies, has always been a source of wonder and intrigue. Among the many milestones in our quest to understand the cosmos, the discovery of exoplanets stands out as a monumental achievement. One such discovery that has captured the imagination of scientists and the public alike is that of K2-18b, a super-Earth located in the habitable zone of its parent star. This chapter delves into the fascinating journey of how astronomers first identified K2-18b and the groundbreaking significance of this discovery.

The Kepler Space Telescope and the K2 Mission

The story of K2-18b begins with the Kepler Space Telescope, an extraordinary instrument launched by NASA in 2009. Kepler's primary mission was to discover Earth-like planets orbiting other stars by detecting the tiny dips in starlight that occur when a planet transits, or passes in front of, its host star. Over its primary mission, Kepler monitored over 150,000 stars and discovered thousands of exoplanet candidates, revolutionizing our understanding of planetary systems.

After the failure of two of its reaction wheels in 2013, Kepler was repurposed for the K2 mission. The K2 mission aimed to continue the search for exoplanets, albeit with a different observational strategy. By utilizing the pressure of sunlight to stabilize its pointing, K2 embarked on a series of "campaigns," each focused on a different region of the sky.

The Discovery of K2-18b

K2-18b was discovered during K2's Campaign 1, which observed a field in the constellation Leo. The planet was identified using the transit method, where astronomers look for periodic dimming in a star's brightness caused by a planet crossing in front of it. The host star, K2-18, is a red dwarf star, smaller and cooler than our Sun, making it an ideal target for detecting habitable zone planets.

The initial detection of K2-18b was made by analyzing the light curves obtained from the K2 mission. These light curves showed periodic dips, indicating the presence of a planet. However, the discovery process involved meticulous data analysis and confirmation through follow-up observations.

Characterizing K2-18b

Upon its discovery, K2-18b was classified as a super-Earth, a type of exoplanet with a mass larger than Earth's but significantly less than that of Neptune or Uranus. K2-18b has an estimated mass of about 8.6 times that of Earth and a radius approximately 2.7 times larger. This places it in a category of planets that are likely to have a thick atmosphere, possibly rich in hydrogen and helium.

One of the most compelling aspects of K2-18b's discovery was its location within the habitable zone of its parent star. The habitable zone, often referred to as the "Goldilocks zone," is the region around a star where conditions might be just right for liquid water to exist on a planet's surface. The presence of liquid water is considered a key ingredient for life as we know it, making planets in this zone prime targets for the search for extraterrestrial life.

Follow-up Observations and the Role of Hubble

The initial discovery of K2-18b was just the beginning. To learn more about this intriguing planet, astronomers conducted follow-up observations using other powerful telescopes. One of the most significant contributions came from the Hubble Space Telescope, which was used to study the atmosphere of K2-18b in greater detail.

In 2019, a team of researchers led by Björn Benneke of the Université de Montréal made a groundbreaking announcement: they had detected water vapor in the atmosphere of K2-18b. This discovery was made possible through Hubble's Wide Field Camera 3, which observed the planet as it transited its star. By analyzing the starlight that passed through K2-18b's atmosphere, the team identified the spectral signature of water vapor, indicating that K2-18b has a substantial atmosphere that could potentially support liquid water.

The Significance of Water Vapor

The detection of water vapor on K2-18b was a monumental achievement in the field of exoplanet research. Water is a fundamental requirement for life as we know it, and its presence on an exoplanet within the habitable zone significantly enhances the planet's potential to support life. The discovery of water vapor on K2-18b thus opened up exciting possibilities for future research and exploration.

Moreover, the detection of water vapor provided valuable insights into the planet's atmospheric composition and structure. It suggested that K2-18b might have a diverse and dynamic climate system, with potential weather patterns and cloud formations. Understanding these atmospheric processes is crucial for assessing the planet's habitability and the potential for life to exist.

Challenges and Future Prospects

While the discovery of K2-18b and the detection of water vapor are exciting milestones, they also present several challenges. One of the primary challenges is the need for more precise measurements and observations to better understand the planet's atmosphere and surface conditions. The exact composition and density of the atmosphere, the presence of other gases, and the planet's surface temperature and pressure remain areas of active research.

Future telescopes and missions, such as the James Webb Space Telescope (JWST) and the European Space Agency's ARIEL mission, are expected to provide more detailed observations of K2-18b and other exoplanets. These advanced instruments will offer higher resolution and sensitivity, allowing scientists to probe deeper into the atmospheres of distant worlds and gather more comprehensive data.

The JWST, in particular, holds great promise for the study of K2-18b. Scheduled for launch in 2021, the JWST is designed to observe the universe in the infrared spectrum, which is ideal for studying the atmospheres of exoplanets. Its capabilities will enable astronomers to analyze the chemical composition, temperature, and pressure of K2-18b's atmosphere with unprecedented precision. This will provide critical information about the planet's potential habitability and the presence of life-supporting conditions.

The Broader Impact of K2-18b's Discovery

The discovery of K2-18b and the detection of water vapor have far-reaching implications for our understanding of the universe and the search for extraterrestrial life. It underscores the diversity of planetary systems and the potential for habitable environments beyond our solar system. K2-18b serves as a reminder that the universe is vast and filled with possibilities, and that we are only beginning to scratch the surface of what lies beyond our cosmic neighborhood.

Furthermore, the discovery of K2-18b highlights the importance of continued investment in space exploration and scientific research. It demonstrates the value of collaborative efforts among scientists, engineers, and space agencies to push the boundaries of knowledge and technology. The search for exoplanets and the study of their atmospheres are not just scientific endeavors; they are a testament to human curiosity and our relentless pursuit of answers to some of the most profound questions about our existence.

The discovery of K2-18b marks a significant milestone in the field of exoplanet research. From its initial detection by the Kepler Space Telescope to the groundbreaking identification of water vapor by the Hubble Space Telescope, K2-18b has captivated the scientific community and the public with its potential to harbor life. As we continue to explore this fascinating world and others like it, we move closer to answering the age-old question: Are we alone in the universe?

In the chapters that follow, we will delve deeper into the characteristics of K2-18b, its atmospheric phenomena, and the broader implications of its discovery. Join us on this journey as we explore "The Steam Planet" and uncover the secrets of life and atmosphere on K2-18b.

2. THE SCIENCE OF EXOPLANETS

Understanding the methods and technologies that enable us to discover and study distant worlds.

The discovery and study of exoplanets—planets orbiting stars outside our solar system—represent one of the most groundbreaking advancements in modern astronomy. This field has transformed our understanding of the universe and opened up new possibilities for discovering habitable worlds and extraterrestrial life. The science of exoplanets is a complex and interdisciplinary field that leverages a variety of methods and technologies to detect and analyze these distant worlds. This chapter delves into the key methods and technologies that have enabled astronomers to uncover the secrets of exoplanets, bringing us closer to understanding planets like K2-18b.

The Transit Method

The transit method is one of the most successful techniques for discovering exoplanets. This method involves monitoring the brightness of a star over time to detect periodic dimming caused by a planet passing in front of it. When a planet transits its star, it blocks a small fraction of the star's light, leading to a temporary dip in brightness. By analyzing these light curves, astronomers can infer the presence of a planet and determine its size and orbital period.

NASA's Kepler Space Telescope, launched in 2009, revolutionized exoplanet discovery using the transit method. Kepler monitored over

150,000 stars in a single patch of sky, searching for the telltale signs of transiting planets. The mission was highly successful, identifying over 2,600 confirmed exoplanets and thousands of additional candidates. This treasure trove of data provided the foundation for many subsequent discoveries and studies, including the identification of K2-18b.

The transit method's effectiveness relies on precise photometry—the measurement of light intensity. Advanced photometric techniques and instruments, both ground-based and space-based, are crucial for detecting the minute changes in brightness caused by transiting exoplanets. Instruments such as the Transiting Exoplanet Survey Satellite (TESS) continue Kepler's legacy, scanning the entire sky to find new exoplanets.

The Radial Velocity Method

The radial velocity method, also known as the Doppler method, detects exoplanets by measuring the wobble in a star's motion caused by the gravitational pull of an orbiting planet. As a planet orbits, it exerts a gravitational force on its host star, causing the star to move in a small orbit around the system's center of mass. This motion induces a periodic shift in the star's spectral lines due to the Doppler effect.

When the star moves toward us, its light is blueshifted (shortened wavelengths), and when it moves away, its light is redshifted (lengthened wavelengths). By analyzing these shifts, astronomers can determine the presence of a planet and estimate its mass and orbital characteristics. The radial velocity method has been instrumental in discovering many of the first known exoplanets, particularly those in close orbits around their stars.

Ground-based observatories equipped with high-resolution spectrographs, such as the HARPS (High Accuracy Radial velocity Planet Searcher) instrument in Chile, have been pivotal in radial velocity measurements. These instruments can detect velocity changes as small as a few meters per second, enabling the discovery of planets with relatively low masses, including super-Earths and Earth-sized planets.

Direct Imaging

Direct imaging involves capturing actual images of exoplanets by blocking out the overwhelming light of their host stars. This method is challenging due to the vast difference in brightness between stars and their planets, as well as the small angular separation between them. However, advancements in adaptive optics and coronagraphy have made direct imaging increasingly feasible.

Adaptive optics systems correct for the blurring effects of Earth's atmosphere in real-time, allowing telescopes to achieve much sharper images. Coronagraphs and starshades are specialized instruments that block out a star's light, making it possible to observe the faint light reflected or emitted by orbiting planets.

Direct imaging provides valuable information about exoplanets, including their atmospheres, compositions, and even weather patterns. The Gemini Planet Imager and the European Southern Observatory's SPHERE instrument are among the leading tools for direct imaging, enabling astronomers to capture detailed images of distant planetary systems.

Gravitational Microlensing

Gravitational microlensing exploits the gravitational field of a massive object, such as a star, to magnify the light from a more distant background star. If a planet orbits the foreground star, it can create an additional, distinct lensing effect, causing a temporary increase in the observed brightness of the background star. By analyzing these light curves, astronomers can detect the presence of the planet and estimate its mass and distance from the host star.

Microlensing is particularly effective for finding planets at greater distances from their stars and for detecting planets in regions of the galaxy that are difficult to study with other methods. The OGLE (Optical Gravitational Lensing Experiment) and MOA (Microlensing Observations in Astrophysics) collaborations have been successful in discovering numerous exoplanets through this technique.

Astrometry

Astrometry is the precise measurement of a star's position in the sky. As a planet orbits its star, it causes the star to move in a small orbit, creating a periodic shift in the star's position. By measuring these tiny shifts, astronomers can infer the presence of a planet and determine its mass and orbital characteristics.

The European Space Agency's Gaia mission is revolutionizing astrometry with its unprecedented precision. Gaia is mapping the positions and motions of over a billion stars in the Milky Way, providing invaluable data for detecting exoplanets and understanding the dynamics of our galaxy.

Atmospheric Characterization

Once an exoplanet is discovered, the next step is often to characterize its atmosphere. This involves analyzing the light that passes through or is emitted by the planet's atmosphere to determine its composition, temperature, and other properties. There are several methods for atmospheric characterization:

Transmission Spectroscopy: During a transit, some of the starlight passes through the planet's atmosphere. By comparing the spectrum of the star's light during and outside the transit, astronomers can identify the absorption features of different molecules in the planet's atmosphere.

Emission Spectroscopy: When a planet passes behind its star (secondary eclipse), the light from the planet is blocked. By comparing the combined light of the star and planet before and after the eclipse, astronomers can isolate the planet's emitted light and analyze its spectrum.

Phase Curve Observations: By observing the changes in a planet's brightness over its orbit, astronomers can infer details about its atmospheric composition and temperature distribution. These observations can reveal information about the planet's climate, weather patterns, and even potential signs of habitability.

The Hubble Space Telescope and the upcoming James Webb Space Telescope (JWST) are among the most powerful tools for atmospheric characterization. These instruments allow astronomers to probe the atmospheres of distant exoplanets with high precision, providing critical insights into their potential for habitability.

The Role of Future Missions

The study of exoplanets is poised for significant advancements with upcoming missions and next-generation telescopes. The James Webb Space Telescope, scheduled for launch in 2021, will revolutionize exoplanet research with its advanced infrared capabilities. JWST will provide unparalleled sensitivity and resolution, enabling detailed studies of exoplanet atmospheres, compositions, and temperatures.

The European Space Agency's ARIEL mission, set to launch in the late 2020s, will focus on the atmospheric characterization of exoplanets. ARIEL will observe a large and diverse sample of exoplanets, providing a comprehensive understanding of their atmospheres and helping to answer fundamental questions about planetary formation and evolution.

The Wide Field Infrared Survey Telescope (WFIRST), another future NASA mission, will employ microlensing and direct imaging techniques to discover and study a wide range of exoplanets, from Earth-sized to gas giants. WFIRST will also investigate the prevalence of potentially habitable worlds in our galaxy.

Ground-based observatories, such as the Extremely Large Telescope (ELT) and the Thirty Meter Telescope (TMT), will complement these space missions with their enormous light-gathering capabilities and advanced instrumentation. These telescopes will enable high-resolution imaging and spectroscopy of exoplanets, pushing the boundaries of what we can learn about distant worlds.

The Broader Implications

The science of exoplanets has profound implications for our understanding of the universe and our place within it. Discovering and studying distant worlds expands our knowledge of planetary systems, their formation, and their diversity. It challenges our assumptions about what makes a planet habitable and offers new perspectives on the conditions necessary for life.

The search for exoplanets also inspires technological innovation and international collaboration. It drives the development of cutting-edge instruments and techniques, pushing the boundaries of what is possible in astronomy and other fields. The pursuit of knowledge about distant worlds unites scientists, engineers, and enthusiasts from around the globe, fostering a spirit of exploration and discovery.

Ultimately, the study of exoplanets brings us closer to answering one of humanity's most profound questions: Are we alone in the universe? By exploring the diverse and dynamic worlds beyond our solar system, we inch closer to finding potential signs of life and understanding the broader context of life in the cosmos.

The science of exoplanets is a testament to human curiosity and ingenuity. Through a combination of innovative methods and advanced technologies, astronomers have made remarkable strides in discovering and studying distant worlds. From the transit method to radial velocity measurements, direct imaging, and atmospheric characterization, each technique contributes to our growing understanding of exoplanets and their potential for habitability.

As we continue to explore the cosmos with next-generation telescopes and missions, the future of exoplanet research looks brighter than ever. The discoveries we make will not only enhance our knowledge of the universe but also inspire future generations to reach for the stars and seek answers to the mysteries of existence.

In the chapters that follow, we will delve deeper into the specific characteristics and atmospheric phenomena of K2-18b, one of the most intriguing exoplanets discovered to date. Join us as we continue our journey through "The Steam Planet" and uncover the secrets of life and atmosphere on K2-18b.

3. A RED DWARF'S COMPANION

Exploring K2-18b's host star, K2-18, and its influence on the planet.

The discovery of K2-18b, a super-Earth exoplanet located in the habitable zone of its star, has fascinated astronomers and the public alike. While much of the excitement revolves around the planet itself and its potential to harbor life, understanding the characteristics and influence of its host star, K2-18, is equally crucial. In this chapter, we delve into the nature of K2-18, a red dwarf star, and examine how its properties and behavior shape the environment and habitability of K2-18b.

Understanding Red Dwarfs

Red dwarfs, also known as M-dwarfs, are the most common type of star in the Milky Way galaxy. These stars are smaller and cooler than our Sun, with masses ranging from about 0.08 to 0.6 times the mass of the Sun. Due to their low mass and temperature, red dwarfs are much less luminous, emitting a faint red light that is often invisible to the naked eye. Despite their dimness, red dwarfs have incredibly long lifespans, often lasting for tens to hundreds of billions of years, far exceeding the lifespan of more massive stars like the Sun.

The prevalence of red dwarfs in the galaxy makes them prime targets in the search for exoplanets. Their small size and low luminosity enhance the detectability of transiting planets, as the relative dip in brightness caused by a planet passing in front of the star is more pronounced. Additionally, the habitable zones of red dwarfs are much closer to the star compared to stars like the Sun, making it easier to detect and study potentially habitable planets.

K2-18: Characteristics and Observations

K2-18 is a red dwarf star located approximately 124 light-years away in the constellation Leo. With a mass of about 0.41 times that of the Sun and a radius roughly 0.4 times the Sun's radius, K2-18 is a relatively typical example of an M-dwarf. Its surface temperature is around 3,450 Kelvin, significantly cooler than the Sun's 5,778 Kelvin, resulting in a much lower luminosity. K2-18's faintness and proximity make it an excellent candidate for exoplanet studies, particularly with the methods of transit photometry and radial velocity measurements.

The Kepler Space Telescope, during its K2 mission, first identified K2-18b by detecting periodic dips in the star's brightness. These dips indicated the presence of a planet transiting the star, leading to the discovery of K2-18b. Subsequent observations with ground-based telescopes, such as the HARPS spectrograph at the La Silla Observatory in Chile, confirmed the planet's existence and provided additional details about its mass and orbital parameters.

The Influence of K2-18 on K2-18b

K2-18's properties and behavior play a crucial role in shaping the environment of K2-18b. Understanding these influences is essential for assessing the planet's habitability and potential to support life.

Stellar Radiation and Temperature: The habitable zone around a star is the region where conditions are suitable for liquid water to exist on a planet's surface. For K2-18, this zone is much closer to the star compared to the Sun's habitable zone, due to the star's lower luminosity. K2-18b orbits within this habitable zone at a distance of approximately 0.142 AU (astronomical units), where it receives about 5% more stellar radiation than Earth receives from the Sun. This proximity results in temperatures that could allow for the presence of liquid water, a critical factor for life as we know it.

Tidal Locking: Given its close orbit, K2-18b is likely tidally locked to its star, meaning one side of the planet always faces the star while the other side is in perpetual darkness. This creates a stark contrast between the day side and night side temperatures. The day side could be much warmer due to continuous stellar irradiation, while the night side remains colder. Tidal locking can also lead to complex atmospheric dynamics, with strong winds and heat circulation potentially redistributing heat across the planet.

Stellar Activity and Flares: Red dwarfs, especially younger ones, are known for their high levels of stellar activity, including flares and coronal mass ejections. These energetic events can release intense bursts of radiation and charged particles, which can have significant effects on a planet's atmosphere and potential habitability. High-energy radiation, such as ultraviolet and X-rays, can strip away a planet's atmosphere over time, reducing its ability to retain heat and support life. However, the level of stellar activity varies among red dwarfs, and long-term observations of K2-18 are necessary to assess its activity levels and potential impacts on K2-18b.

Magnetic Fields: The interaction between a planet's magnetic field and its host star's magnetic activity is another important factor. A strong planetary magnetic field can protect the atmosphere from stellar wind and radiation, much like Earth's magnetosphere shields us from the solar wind. Understanding whether K2-18b has a magnetic field and how it interacts with K2-18's magnetic activity is crucial for evaluating the planet's atmospheric retention and habitability.

Atmospheric Composition and Climate

The atmosphere of K2-18b plays a critical role in determining its climate and potential for supporting life. The detection of water vapor in K2-18b's atmosphere by the Hubble Space Telescope was a groundbreaking discovery, indicating that the planet has a substantial atmosphere that could support liquid water under the right conditions.

Water Vapor and Greenhouse Effect: Water vapor is a potent greenhouse gas, capable of trapping heat and warming the planet's surface. The presence of water vapor suggests that K2-18b could experience a significant greenhouse effect, potentially creating a stable and temperate climate conducive to liquid water. The extent of this effect depends on the atmospheric composition, pressure, and other greenhouse gases present.

Cloud Formation and Weather Patterns: The detection of water vapor also implies the possibility of cloud formation and dynamic weather patterns. Clouds can reflect incoming stellar radiation, cooling the surface, or trap heat, depending on their composition and altitude. The interplay between cloud cover and greenhouse gases is crucial for determining the planet's climate and habitability.

Atmospheric Composition: To fully understand K2-18b's atmosphere, scientists must identify other constituents, such as carbon dioxide, methane, and nitrogen. These gases can provide insights into the planet's geological activity, potential biosignatures, and overall climate stability. Future observations with more advanced telescopes, such as the James Webb Space Telescope (JWST), will be instrumental in characterizing the atmospheric composition of K2-18b in greater detail.

The Future of K2-18b Research

The discovery of K2-18b and the detection of water vapor in its atmosphere have set the stage for future research aimed at understanding its potential habitability. Several upcoming missions and observational campaigns will provide valuable data to advance our knowledge of this intriguing exoplanet and its red dwarf host star.

James Webb Space Telescope (JWST): Scheduled for launch in 2021, JWST will be a game-changer in exoplanet research. With its advanced infrared capabilities, JWST will be able to probe the atmospheres of distant worlds with unprecedented precision. Observations of K2-18b with JWST will help determine the detailed atmospheric composition, temperature, and pressure, providing critical insights into its potential habitability.

ARIEL Mission: The European Space Agency's ARIEL (Atmospheric Remote-sensing Infrared Exoplanet Large-survey) mission, set to launch in the late 2020s, will focus on the atmospheric characterization of exoplanets. ARIEL will observe a large and diverse sample of exoplanets, including K2-18b, to understand their atmospheres and evolutionary processes.

Ground-Based Observatories: Next-generation ground-based telescopes, such as the Extremely Large Telescope (ELT) and the Thirty Meter Telescope (TMT), will complement space missions by providing high-resolution spectroscopy and imaging. These observatories will enhance our ability to study K2-18 and its planetary system, offering deeper insights into the star-planet interactions and the broader context of red dwarf systems.

Stellar Activity Monitoring: Long-term monitoring of K2-18's stellar activity is essential to assess the impact of flares and other energetic events on K2-18b's atmosphere and habitability. Understanding the variability and intensity of K2-18's activity will help scientists evaluate the potential risks and protective mechanisms for life on the planet.

K2-18b's status as a super-Earth in the habitable zone of a red dwarf star makes it a compelling target for exoplanet research. The characteristics and behavior of its host star, K2-18, play a pivotal role in shaping the planet's environment and potential for supporting life. From stellar radiation and tidal locking to stellar activity and atmospheric composition, the interplay between K2-18 and K2-18b offers a fascinating glimpse into the complexities of planetary habitability.

As we continue to explore K2-18b with advanced telescopes and missions, we move closer to understanding the conditions necessary for life beyond our solar system. The study of K2-18 and its planetary companion not only deepens our knowledge of red dwarf systems but also inspires us to keep searching for signs of life in the vast and varied universe.

In the chapters that follow, we will delve into the specific characteristics and atmospheric phenomena of K2-18b, exploring its potential for habitability and the intriguing possibility of life on a distant world. Join us as we continue our journey through "The Steam Planet" and uncover the secrets of life and atmosphere on K2-18b.

4. SUPER-EARTHS

A New Class of Planets: Defining super-Earths and what makes K2-18b unique among them.

The discovery of exoplanets has revolutionized our understanding of the universe and our place within it. Among the diverse array of exoplanets identified, super-Earths have emerged as a particularly intriguing class. These planets, which have masses between those of Earth and Neptune, offer a tantalizing glimpse into the potential variety of rocky worlds. K2-18b, a super-Earth located in the habitable zone of its host star, K2-18, stands out as a unique and compelling member of this class. This chapter explores the defining characteristics of super-Earths, examines what sets K2-18b apart, and delves into the broader implications of these fascinating worlds.

Defining Super-Earths

Super-Earths are exoplanets with masses ranging from approximately 1 to 10 Earth masses. This definition is primarily based on mass, as these planets can vary widely in their composition, size, and atmospheric properties. Super-Earths occupy a middle ground between Earth-like terrestrial planets and gas giants like Neptune and Jupiter. Despite their name, super-Earths are not necessarily similar to Earth in terms of habitability or other characteristics; the term simply denotes their mass range.

Characteristics of Super-Earths

Composition and Structure: The composition of super-Earths can vary significantly. Some may be rocky, similar to Earth, while others could have substantial amounts of gas or water. The internal structure of a super-Earth depends on its formation history and composition. Models suggest that these planets could have a dense rocky core surrounded by layers of metal, silicate, water, and gas in varying proportions.

Atmospheric Properties: The atmospheres of super-Earths are diverse, ranging from thin, Earth-like atmospheres to thick, hydrogen-rich envelopes. The atmospheric composition and pressure influence a planet's surface conditions, including temperature and potential habitability. The presence of greenhouse gases, such as carbon dioxide and water vapor, can significantly affect the climate and weather patterns on these planets.

Orbital Characteristics: Super-Earths are found in a variety of orbital configurations, from close-in planets with short orbital periods to those in wider orbits. Their proximity to their host stars can affect their atmospheric retention, surface temperatures, and potential for tidal locking, where one side of the planet always faces the star.

Potential for Habitability: The habitability of super-Earths depends on several factors, including their distance from the host star, atmospheric composition, and geological activity. Those located in the habitable zone, where conditions could allow for liquid water, are of particular interest. However, the mere presence in the habitable zone does not guarantee habitability, as atmospheric and surface conditions play crucial roles.

K2-18b: A Unique Super-Earth

K2-18b, discovered by the Kepler Space Telescope during its K2 mission, has captivated astronomers due to its location in the habitable zone of its host star, K2-18, and its potential for hosting life. This super-Earth is approximately 2.6 times the radius of Earth and about 8.6 times its mass, placing it firmly within the super-Earth category. Several factors contribute to K2-18b's uniqueness among super-Earths:

Location in the Habitable Zone: One of the most significant aspects of K2-18b is its position within the habitable zone of K2-18, a red dwarf star. The habitable zone is the region around a star where conditions could potentially allow for liquid water to exist on a planet's surface. K2-18b's location in this zone raises the possibility of a temperate climate and the presence of liquid water, both crucial for life as we know it.

Atmospheric Composition: Observations using the Hubble Space Telescope have detected water vapor in K2-18b's atmosphere, a groundbreaking discovery that suggests the planet has a substantial atmosphere capable of supporting liquid water. The detection of water vapor indicates that K2-18b's atmosphere may be similar to that of Earth in terms of its ability to sustain water, albeit under potentially different conditions due to the planet's greater mass and radius.

Stellar Environment: K2-18b orbits a red dwarf star, which significantly influences its environment and potential habitability. Red dwarfs are known for their high levels of stellar activity, including flares and coronal mass ejections. These energetic events can impact a planet's atmosphere, stripping it away over time or causing significant climatic variations. Understanding the level of stellar activity of K2-18 is crucial for assessing the long-term habitability of K2-18b.

Potential for Tidal Locking: Due to its close orbit, K2-18b is likely tidally locked, meaning one side of the planet always faces the star while the other side is in perpetual darkness. This creates a stark contrast between the day side and night side temperatures, leading to complex atmospheric dynamics. The potential for a stable climate with a temperate zone between the day and night sides could exist, depending on the efficiency of heat redistribution by the atmosphere.

The Broader Implications of Super-Earths

The study of super-Earths like K2-18b has far-reaching implications for our understanding of planetary systems and the potential for life beyond our solar system. These planets provide unique opportunities to explore the diversity of planetary compositions, atmospheres, and potential habitability.

Planetary Formation and Evolution: Super-Earths challenge existing models of planetary formation and evolution. Their intermediate size and mass suggest different formation processes compared to smaller terrestrial planets and larger gas giants. Studying super-Earths can provide insights into the various pathways of planetary development and the conditions that lead to their formation.

Atmospheric Studies: The diverse atmospheres of super-Earths offer a natural laboratory for studying atmospheric processes and climate systems. By comparing the atmospheric properties of different super-Earths, scientists can develop a better understanding of how atmospheres evolve, interact with stellar radiation, and influence surface conditions.

Habitability and Life: The potential habitability of super-Earths extends the search for life beyond our solar system. Planets like K2-18b, located in the habitable zone with detectable water vapor, provide compelling targets for future investigations of biosignatures—chemical indicators of life. The discovery of life on a super-Earth would have profound implications for our understanding of life's prevalence and diversity in the universe.

Technological Advancements: The study of super-Earths drives technological advancements in astronomy and planetary science. The development of more sensitive instruments, such as the James Webb Space Telescope (JWST) and next-generation ground-based observatories, enables more detailed observations of exoplanet atmospheres and surfaces. These technologies not only enhance our ability to study super-Earths but also contribute to broader scientific and technological progress.

Future Research and Exploration

The exploration of super-Earths, including K2-18b, is poised for significant advancements with upcoming missions and observational campaigns. Several key initiatives will provide valuable data to deepen our understanding of these intriguing worlds:

James Webb Space Telescope (JWST): Scheduled for launch in 2021, JWST will revolutionize exoplanet research with its advanced infrared capabilities. JWST will be able to probe the atmospheres of super-Earths like K2-18b with unprecedented precision, providing detailed information about their composition, temperature, and potential biosignatures.

ARIEL Mission: The European Space Agency's ARIEL (Atmospheric Remote-sensing Infrared Exoplanet Large-survey) mission, set to launch in the late 2020s, will focus on the atmospheric characterization of a large and diverse sample of exoplanets, including super-Earths. ARIEL's observations will enhance our understanding of the atmospheric diversity and evolutionary processes of these planets.

Ground-Based Observatories: Next-generation ground-based telescopes, such as the Extremely Large Telescope (ELT) and the Thirty Meter Telescope (TMT), will complement space missions by providing high-resolution spectroscopy and imaging of super-Earths. These observatories will enable detailed studies of exoplanetary atmospheres and surface conditions, furthering our knowledge of their habitability.

Long-Term Monitoring: Continuous monitoring of super-Earths and their host stars is essential for understanding their dynamic environments and potential habitability. Long-term observations can reveal changes in atmospheric composition, stellar activity, and other factors that influence the habitability of these planets.

Super-Earths represent a fascinating and diverse class of exoplanets that challenge our understanding of planetary systems and the potential for life beyond Earth. K2-18b, with its location in the habitable zone and detectable water vapor in its atmosphere, stands out as a unique and compelling example of a super-Earth with potential habitability. The study of K2-18b and other super-Earths not only enhances our knowledge of planetary formation, atmospheric processes, and habitability but also inspires technological advancements and future exploration.

As we continue to explore the universe with advanced telescopes and missions, the discoveries we make about super-Earths will bring us closer to answering fundamental questions about the prevalence and diversity of life in the cosmos. The journey to understand these intriguing worlds is just beginning, and the insights we gain will undoubtedly shape our understanding of the universe and our place within it.

5. ATMOSPHERIC REVELATIONS

How scientists detected water vapor in K2-18b's atmosphere and what it means.

The quest to find life beyond Earth has taken a significant step forward with the discovery of water vapor in the atmosphere of K2-18b, an exoplanet located about 124 light-years away in the constellation Leo. This remarkable finding, made using the Hubble Space Telescope, has captivated astronomers and the public alike, bringing us closer to understanding the conditions that might support life on other worlds. This chapter delves into the scientific journey that led to this discovery, the techniques used to detect water vapor, and the profound implications this has for the study of exoplanets and the search for extraterrestrial life.

Discovering K2-18b: A New Hope

K2-18b was first discovered in 2015 by NASA's Kepler spacecraft during its K2 mission. The planet orbits within the habitable zone of its host star, a region where temperatures could allow for liquid water to exist on a planet's surface. K2-18b is classified as a super-Earth, a type of exoplanet with a mass larger than Earth's but smaller than that of Uranus or Neptune. With a radius about 2.7 times that of Earth and a mass approximately eight times greater, K2-18b quickly became a subject of interest for astronomers looking for potentially habitable exoplanets.

The Role of the Hubble Space Telescope

The Hubble Space Telescope has been instrumental in expanding our understanding of exoplanet atmospheres. Launched in 1990, Hubble has provided high-resolution data that have been crucial in studying distant planets. For K2-18b, Hubble's Wide Field Camera 3 (WFC3) was used to observe the planet as it transited its host star. During a transit, some of the starlight passes through the planet's atmosphere, allowing scientists to analyze the absorption features in the light spectrum and infer the atmospheric composition.

Detecting Water Vapor: The Methodology

The detection of water vapor in K2-18b's atmosphere involved a sophisticated analysis of the light spectra obtained during the planet's transits. Here's a step-by-step breakdown of how scientists achieved this groundbreaking discovery:

Observation of Transits: Hubble observed K2-18b as it passed in front of its host star. This event, known as a transit, caused a slight dimming of the star's light, which Hubble could measure with great precision.

Spectral Analysis: As the starlight filtered through K2-18b's atmosphere, different molecules absorbed specific wavelengths of light. By analyzing the resulting spectrum, scientists could identify the unique signatures of various gases.

Data Reduction: The raw data from Hubble required extensive processing to remove noise and other distortions. Advanced algorithms were employed to extract the clean spectra needed for analysis.

Modeling Atmospheric Composition: Using the cleaned spectra, researchers applied atmospheric models to simulate the conditions on K2-18b. These models incorporated various potential atmospheric compositions to match the observed spectral features.

Identification of Water Vapor: Among the different models, the one that best fit the observed data included a significant amount of water vapor. The spectral signature of water was evident, confirming its presence in the atmosphere of K2-18b.

Implications for Habitability

The discovery of water vapor on K2-18b has profound implications for the study of habitability outside our solar system. Water is a fundamental ingredient for life as we know it, and its presence on K2-18b raises the possibility that this exoplanet could harbor life. However, several factors need to be considered to assess the planet's true habitability.

Atmospheric Conditions

The detection of water vapor indicates that K2-18b has a substantial atmosphere, but the overall composition and pressure of this atmosphere are critical in determining its suitability for life. If the atmosphere is too thick or composed of gases that are toxic to life as we know it, the planet might not be habitable despite the presence of water vapor.

Temperature and Climate

K2-18b lies within the habitable zone of its star, where temperatures could allow for liquid water. However, the exact surface temperature depends on the greenhouse effect caused by the planet's atmosphere. If the greenhouse gases trap too much heat, the surface could become too hot for life. Conversely, if the atmosphere does not trap enough heat, the planet could be too cold.

Surface Conditions

The presence of water vapor in the atmosphere suggests that K2-18b might have liquid water on its surface, but this is not guaranteed. The planet's surface could be covered with ice or clouds, or the water might exist only in vapor form. Future observations and studies are needed to determine the exact state of water on K2-18b.

Potential for Life

While the discovery of water vapor is a promising sign, it does not necessarily mean that K2-18b hosts life. Life requires a complex interplay of various factors, including the right chemical ingredients, energy sources, and stable environmental conditions. The presence of water vapor is one piece of the puzzle, but it is not conclusive evidence of life.

The Future of Exoplanet Research

The discovery of water vapor in K2-18b's atmosphere represents a significant milestone in exoplanet research, but it also highlights the need for more advanced observational tools and techniques. Future missions and telescopes will play a crucial role in furthering our understanding of exoplanets and their potential to host life.

The James Webb Space Telescope (JWST)

Scheduled for launch in 2021, the James Webb Space Telescope (JWST) is expected to revolutionize the study of exoplanet atmospheres. With its larger mirror and advanced instruments, JWST will be able to observe exoplanets with unprecedented sensitivity and detail. It will provide more precise measurements of atmospheric compositions and temperatures, offering deeper insights into the habitability of planets like K2-18b.

Ground-Based Observatories

Next-generation ground-based observatories, such as the Extremely Large Telescope (ELT) and the Giant Magellan Telescope (GMT), will complement space-based observations. Equipped with advanced adaptive optics, these telescopes will be able to directly image exoplanets and analyze their atmospheres. These facilities will help confirm the presence of water and other key molecules on distant worlds.

Space Missions

In addition to telescopes, future space missions dedicated to the study of exoplanets will enhance our capabilities. Missions like the Transiting Exoplanet Survey Satellite (TESS) and the PLAnetary Transits and Oscillations of stars (PLATO) mission will continue to discover new exoplanets and provide valuable data for atmospheric studies.

The detection of water vapor in the atmosphere of K2-18b is a landmark achievement in the field of exoplanet research. It demonstrates the incredible progress that has been made in the search for habitable worlds beyond our solar system. While K2-18b might not be the definitive answer to the question of extraterrestrial life, it offers a tantalizing glimpse of what lies beyond. The methods and technologies developed to study K2-18b will pave the way for future discoveries, bringing us closer to finding another Earth-like planet and, perhaps, life beyond our own. As we continue to explore the cosmos, K2-18b serves as a beacon of possibility and a reminder of the vast potential that the universe holds.

6. THE STEAMY SKIES OF K2-18B

Diving into the composition and characteristics of the planet's steamy atmosphere.

K2-18b, a captivating exoplanet located about 124 light-years away in the constellation Leo, has emerged as a focal point in the study of potentially habitable worlds beyond our solar system. This super-Earth, orbiting within the habitable zone of its red dwarf star, has fascinated astronomers since its discovery in 2015 by NASA's Kepler spacecraft. One of the most groundbreaking revelations about K2-18b is the presence of water vapor in its atmosphere, a discovery that has profound implications for our understanding of planetary atmospheres and the potential for life elsewhere in the universe. This chapter explores the intricate composition and characteristics of K2-18b's steamy atmosphere, delving into the scientific methods that unveiled its secrets and discussing what these findings mean for the broader field of exoplanet research.

Unveiling the Atmosphere: The Role of Advanced Observations

The journey to uncover the atmospheric composition of K2-18b began with observations using the Hubble Space Telescope. By analyzing the light spectra during the planet's transits across its host star, scientists were able to infer the presence of various atmospheric constituents. These observations were particularly focused on the absorption features in the light spectrum, which provide critical clues about the types and quantities of gases present.

The Transit Method

The transit method, a cornerstone in the study of exoplanets, involves measuring the dimming of a star's light as a planet passes in front of it. This dimming effect, although slight, can reveal a wealth of information about the planet, including its size, orbital period, and atmospheric properties. For K2-18b, each transit provided an opportunity to capture starlight that had passed through its atmosphere, offering a direct window into its composition.

Spectroscopy and Atmospheric Analysis

Using the Wide Field Camera 3 (WFC3) on Hubble, astronomers obtained high-resolution spectra of K2-18b during multiple transits. The spectral data indicated the presence of water vapor, identified by its distinct absorption lines. This discovery marked the first time water vapor was detected in the atmosphere of a super-Earth within the habitable zone, making K2-18b a prime candidate for further atmospheric studies.

Composition of K2-18b's Atmosphere

The detection of water vapor in K2-18b's atmosphere was a monumental step, but it also raised numerous questions about the overall atmospheric composition and its implications for habitability. To understand the steamy skies of K2-18b, scientists employed advanced modeling techniques to interpret the spectral data and estimate the relative abundances of various gases.

Water Vapor

Water vapor is a critical component of K2-18b's atmosphere. Its presence suggests that the planet might have a significant amount of water, either in the form of vapor, clouds, or even liquid on its surface. The exact concentration of water vapor is still a subject of ongoing research, but its detection alone provides a strong indication that K2-18b has a dynamic and possibly complex atmospheric system.

Hydrogen and Helium

In addition to water vapor, the atmosphere of K2-18b is likely dominated by hydrogen and helium, similar to the gas giants in our solar system. This composition is typical for planets of its size and mass, which are often referred to as mini-Neptunes. The abundance of these lighter gases contributes to a thick, expansive atmosphere that can significantly affect the planet's climate and surface conditions.

Potential for Other Molecules

While hydrogen, helium, and water vapor are the primary constituents identified so far, there is potential for other molecules to exist in K2-18b's atmosphere. Methane (CH_4), ammonia (NH_3), and other hydrocarbons could also be present, each contributing to the overall chemical complexity. Future observations with more advanced telescopes, such as the James Webb Space Telescope (JWST), are expected to provide deeper insights into these possibilities.

Atmospheric Characteristics

Understanding the composition of K2-18b's atmosphere is only part of the story. To fully grasp the nature of this distant world, scientists must also explore its atmospheric characteristics, including temperature, pressure, weather patterns, and potential climate dynamics.

Temperature and Climate

K2-18b's location within the habitable zone suggests that it could have a range of temperatures suitable for liquid water. However, the actual surface temperature depends on several factors, including the atmospheric greenhouse effect and the planet's albedo (reflectivity). Models predict that the greenhouse gases in K2-18b's atmosphere could create a warming effect, potentially leading to a temperate climate on the surface. The extent of this warming and its impact on potential habitability are key areas of ongoing research.

Atmospheric Pressure

The atmospheric pressure on K2-18b is another crucial factor in determining its habitability. A thick atmosphere, rich in hydrogen and helium, could result in high surface pressures, which might pose challenges for life as we know it. Conversely, if the atmosphere is not too dense, conditions could be more favorable. Current models suggest a range of possible pressures, and future observations will be necessary to narrow down these estimates.

Weather and Cloud Formation

The presence of water vapor raises the possibility of weather phenomena on K2-18b, including cloud formation and precipitation. Clouds composed of water droplets or ice crystals could significantly influence the planet's climate by reflecting sunlight and trapping heat. Studying these processes on K2-18b could provide valuable insights into the atmospheric dynamics of super-Earths and mini-Neptunes.

Implications for Habitability

The discovery of water vapor in K2-18b's atmosphere has sparked considerable excitement about its potential habitability. While this finding is a promising indicator, it is only one piece of a complex puzzle. To assess the true potential for life, scientists must consider a multitude of factors beyond atmospheric composition.

Surface Conditions

For K2-18b to be habitable, it must have stable surface conditions that allow for the presence of liquid water. This depends not only on atmospheric composition and pressure but also on the planet's geophysical properties, such as its rotation rate, axial tilt, and geothermal activity. Understanding these factors is critical for building a comprehensive picture of K2-18b's habitability.

Energy Sources

Life requires energy, and for K2-18b, the primary source would be its host star. The red dwarf star K2-18 emits less energy than our Sun, but it also produces significant amounts of ultraviolet and X-ray radiation, which could impact the planet's atmosphere and surface conditions. The balance between beneficial and harmful radiation is a key consideration in evaluating K2-18b's potential to support life.

Chemical Composition and Biochemistry

The presence of water vapor suggests that K2-18b has the potential to support some form of water-based chemistry. However, the availability of other essential elements and compounds, such as carbon, nitrogen, and phosphorus, is also crucial. These elements are the building blocks of life as we know it, and their presence or absence could determine the viability of life on K2-18b.

Future Prospects

The study of K2-18b is still in its early stages, and much remains to be discovered about this intriguing exoplanet. The upcoming launch of the James Webb Space Telescope (JWST) promises to revolutionize our understanding by providing more detailed observations of K2-18b's atmosphere. JWST's advanced instruments will be able to detect a wider range of molecules and provide higher-resolution spectra, offering unprecedented insights into the planet's atmospheric composition and dynamics.

Ground-Based Observations

In addition to space-based telescopes, next-generation ground-based observatories, such as the Extremely Large Telescope (ELT) and the Giant Magellan Telescope (GMT), will play a crucial role in the study of K2-18b. Equipped with advanced adaptive optics, these observatories will be capable of directly imaging exoplanets and conducting detailed atmospheric analyses, further enhancing our understanding of K2-18b's steamy skies.

The Search for Life

Ultimately, the study of K2-18b is driven by the fundamental question of whether life exists beyond Earth. While the presence of water vapor is a significant milestone, it is only the beginning. The search for biosignatures, such as specific gases or molecules associated with biological activity, will be a major focus of future research. Detecting these biosignatures would be a transformative discovery, providing the first direct evidence of life on another world.

The steamy skies of K2-18b offer a tantalizing glimpse into the complexities of exoplanetary atmospheres and the potential for life beyond our solar system. The detection of water vapor is a groundbreaking achievement that highlights the remarkable progress made in the field of exoplanet research. As we continue to explore K2-18b and other distant worlds, the insights gained from these studies will bring us closer to answering one of humanity's most profound questions: Are we alone in the universe? The journey is just beginning, and the discoveries awaiting us promise to be as awe-inspiring as they are transformative.

7. CLIMATE AND WEATHER ON K2-18B

Speculating on the dynamic weather patterns and climate of a steam-filled world.

The discovery of K2-18b, an exoplanet orbiting within the habitable zone of a red dwarf star, has ignited scientific curiosity and public imagination alike. Approximately 124 light-years away in the constellation Leo, K2-18b is a super-Earth with a mass about eight times that of our planet. What makes K2-18b particularly intriguing is the detection of water vapor in its atmosphere, suggesting the potential for a steam-filled world with dynamic weather patterns and a complex climate system. This chapter of "The Steam Planet: Life and Atmosphere on K2-18b" explores the speculative yet scientifically grounded possibilities for the climate and weather on K2-18b.

Understanding K2-18b's Atmosphere

Composition and Structure

K2-18b's atmosphere is primarily composed of hydrogen and helium, typical of many exoplanets. However, the presence of water vapor sets it apart, hinting at more complex weather systems. The atmospheric pressure and temperature likely vary significantly from those on Earth, given the planet's greater mass and different orbital dynamics. These factors contribute to a unique environment where weather patterns could be both fascinating and extreme.

The Role of Water Vapor

Water vapor plays a crucial role in Earth's climate by influencing weather patterns, cloud formation, and precipitation. On K2-18b, water vapor might similarly drive dynamic weather systems. The presence of water vapor in significant amounts suggests the possibility of a hydrological cycle, albeit one that could differ greatly from Earth's due to varying temperatures and atmospheric pressures.

Temperature Gradients and Climate Zones

Equatorial Heat and Polar Chill

K2-18b likely experiences substantial temperature gradients between its equator and poles. The planet's rotation and the energy received from its host star would create distinct climate zones, much like on Earth. However, given K2-18b's proximity to a red dwarf star, its climate zones might be more extreme.

Equatorial Region

The equatorial region of K2-18b would receive the most intense solar radiation, potentially leading to high temperatures and significant atmospheric convection. This could result in strong, persistent storms and cloud formation, driven by the rising warm, moist air.

Polar Regions

In contrast, the polar regions might be significantly cooler. The reduced solar input could result in cold, stable air masses, with less dynamic weather compared to the equator. However, if the planet's atmosphere is thick enough, heat distribution might moderate these extremes somewhat.

Day-Night Temperature Differences

If K2-18b is tidally locked, one side would perpetually face its star while the other remains in darkness. This configuration could create extreme temperature differences between the day and night sides, driving powerful winds and dynamic weather systems as the atmosphere attempts to redistribute heat.

Day Side

The day side of a tidally locked K2-18b would be a hot, illuminated hemisphere where intense solar heating could lead to high temperatures, significant evaporation, and a dense steam-filled atmosphere. Strong thermal gradients could drive robust atmospheric circulation, possibly resulting in violent storms and continuous cloud cover.

Night Side

The night side, shrouded in perpetual darkness, would be much cooler. Without direct solar heating, the atmosphere here could condense, leading to potential precipitation of water vapor. The stark contrast with the day side would generate strong winds as air masses move from the hot, high-pressure day side to the cold, low-pressure night side.

Weather Phenomena on K2-18b

Cloud Formation and Dynamics

Clouds on K2-18b would likely be a dominant feature, influenced by the abundant water vapor. The exact nature of these clouds—whether they are similar to Earth's water clouds or composed of different substances due to varying temperatures and pressures—remains a topic of speculation.

Types of Clouds

Cloud types on K2-18b could range from thick, reflective clouds in the upper atmosphere to dense fog-like layers closer to the surface. These clouds would play a critical role in the planet's albedo, reflecting a portion of the star's light and impacting the overall climate.

Cloud Coverage and Climate Impact

Persistent cloud coverage, especially on the day side of a tidally locked planet, could create a greenhouse effect, trapping heat and raising surface temperatures. Conversely, clouds could also reflect solar radiation, leading to cooling effects. The balance between these opposing forces would significantly influence the climate.

Precipitation Patterns

Precipitation on K2-18b would be driven by its hydrological cycle, with water vapor condensing into clouds and falling as rain or other forms of precipitation. The nature and frequency of this precipitation would depend on the atmospheric conditions.

Rainfall

Rainfall might be common in regions where warm, moist air rises and cools, leading to condensation. Equatorial zones and areas where atmospheric circulation drives moisture-laden air upward would likely see frequent and intense rainstorms.

Other Forms of Precipitation

Depending on temperature variations, precipitation could also occur in other forms. For instance, cooler regions might experience condensation leading to dew or frost, while extremely high-altitude clouds could produce hail-like particles.

Wind Patterns

Wind patterns on K2-18b would be shaped by the planet's rotation, atmospheric pressure differences, and temperature gradients. The stronger these factors, the more dynamic and powerful the winds.

Global Circulation

If K2-18b is tidally locked, the global circulation pattern would feature strong winds flowing from the hot day side to the cold night side. This could result in a continuous, planet-wide wind system, potentially reaching hundreds of kilometers per hour.

Storm Systems

The combination of high winds, moisture, and thermal gradients could lead to frequent and intense storm systems. These storms might resemble terrestrial hurricanes but could be far more massive and long-lasting due to the planet's size and atmospheric dynamics.

Climate Modeling and Predictions

Simulating K2-18b's Climate

Climate models are essential tools for predicting the weather and climate on K2-18b. By inputting data on the planet's size, atmospheric composition, orbital characteristics, and stellar radiation, scientists can simulate various scenarios and predict the likely climate and weather patterns.

Key Variables

Key variables in these models include the planet's albedo, greenhouse gas concentrations, atmospheric pressure, and the distribution of water vapor. Adjusting these variables allows researchers to explore a range of possible climates, from scorching steam worlds to more temperate, Earth-like environments.

Model Limitations

While climate models offer valuable insights, they also have limitations. The lack of direct observational data from K2-18b means that many inputs are based on assumptions or analogs from known planets. As a result, predictions remain speculative and subject to revision as new data becomes available.

Observational Data and Future Missions

Future space missions and telescopes, such as the James Webb Space Telescope (JWST) and the Extremely Large Telescope (ELT), will provide more detailed observations of K2-18b's atmosphere. These observations will refine climate models and improve our understanding of the planet's weather patterns.

Atmospheric Composition Analysis

Spectroscopic data from these missions will reveal the precise composition of K2-18b's atmosphere, identifying key gases and their concentrations. This information is crucial for accurately modeling the greenhouse effect and other climatic processes.

Temperature and Pressure Profiles

Detailed measurements of temperature and pressure profiles across different regions of K2-18b will enhance climate models, allowing for more accurate predictions of weather phenomena. This data will help scientists understand how heat is distributed and how atmospheric dynamics operate on the planet.

Implications for Habitability

Potential for Liquid Water

The presence of water vapor suggests that liquid water might exist on K2-18b, either on its surface or within its atmosphere. Liquid water is a key ingredient for life as we know it, making K2-18b a compelling candidate in the search for extraterrestrial life.

Surface Water

If K2-18b has a rocky surface, liquid water could exist in oceans, lakes, or rivers, particularly in regions where temperatures and pressures are conducive. These bodies of water would play a central role in the planet's climate, influencing weather patterns and atmospheric dynamics.

Atmospheric Water

Even if K2-18b lacks a solid surface, liquid water droplets could form within its atmosphere, contributing to cloud formation and precipitation. This atmospheric water cycle would be crucial for understanding the planet's habitability and potential for supporting life.

Climate Stability

The stability of K2-18b's climate is a significant factor in assessing its habitability. A stable climate with moderate temperatures and consistent water availability would be more conducive to life than one with extreme fluctuations.

Long-Term Climate Trends

Long-term climate trends, influenced by the planet's orbit, rotation, and stellar activity, will determine the persistence of habitable conditions. Understanding these trends is essential for evaluating K2-18b's potential as a home for life.

Potential for Life

While the presence of water vapor and dynamic weather patterns suggests potential habitability, the actual presence of life would depend on numerous factors, including the availability of essential nutrients and the planet's ability to sustain stable, life-supporting conditions over long periods.

Speculating on the climate and weather of K2-18b offers a glimpse into the complexities and wonders of exoplanetary science. This steam-filled world, with its potential for dynamic weather patterns and diverse climate zones, challenges our understanding of planetary atmospheres and their capacity to support life. As future missions and technologies continue to explore K2-18b, we will gain deeper insights into its climate and weather, bringing us closer to answering profound questions about the nature of habitable worlds beyond our solar system. The study of K2-18b not only advances our scientific knowledge but also fuels our imagination, inspiring us to envision the myriad possibilities that exist within the vast expanse of the universe.

8. HABITABILITY BEYOND THE GOLDILOCKS

Zone Examining the factors that could make K2-18b a candidate for habitability.

The search for extraterrestrial life has long fascinated scientists and the public alike. The discovery of K2-18b, a super-Earth orbiting within the habitable zone of its host star, has reignited this interest. Located approximately 124 light-years away in the constellation Leo, K2-18b is notable for its potential to support life. This chapter of "The Steam Planet: Life and Atmosphere on K2-18b" delves into the various factors that could make K2-18b a candidate for habitability, exploring beyond the traditional Goldilocks Zone concept.

The Goldilocks Zone

Defining the Goldilocks Zone

The Goldilocks Zone, or the habitable zone, refers to the region around a star where conditions are just right for liquid water to exist on a planet's surface—neither too hot nor too cold. This concept is foundational in the search for habitable exoplanets, as liquid water is considered essential for life as we know it.

K2-18b in the Goldilocks Zone

K2-18b orbits within the habitable zone of its red dwarf star, K2-18. This positioning suggests that the planet could maintain temperatures conducive to liquid water. However, being in the Goldilocks Zone is just one of many factors that influence a planet's habitability.

Atmospheric Composition

The Presence of Water Vapor

One of the most exciting discoveries about K2-18b is the detection of water vapor in its atmosphere. This finding, made using the Hubble Space Telescope, indicates that the planet has an atmosphere that can support water in its gaseous form. Water vapor is a key component in creating and sustaining a habitable environment.

Other Atmospheric Gases

In addition to water vapor, the presence of other gases such as oxygen, methane, and carbon dioxide can significantly impact a planet's habitability. These gases can provide insights into potential biological processes. For instance, oxygen and methane together can indicate biological activity, as they typically react and degrade in each other's presence without constant replenishment.

Greenhouse Effect

The composition of K2-18b's atmosphere will also determine its greenhouse effect. A robust greenhouse effect can trap heat and maintain surface temperatures conducive to liquid water. However, an excessive greenhouse effect could lead to runaway heating, making the planet uninhabitable.

Surface Conditions

Liquid Water

The potential presence of liquid water on K2-18b's surface is a critical factor for habitability. Liquid water is essential for life as we know it, serving as a solvent for biochemical reactions. If K2-18b has stable bodies of water, it significantly enhances the planet's potential for supporting life.

Geological Activity

Geological activity, such as volcanism and plate tectonics, can play a crucial role in maintaining a planet's habitability. These processes can recycle nutrients, create diverse habitats, and regulate atmospheric composition. Evidence of geological activity on K2-18b would strengthen its candidacy as a habitable world.

Magnetic Field

A magnetic field can protect a planet's atmosphere from being stripped away by stellar winds. This protection is vital for maintaining a stable atmosphere and, by extension, a stable climate. While the presence of a magnetic field on K2-18b is still speculative, its potential existence is a factor in the planet's habitability.

Stellar Environment

Red Dwarf Stars and Habitability

K2-18b orbits a red dwarf star, which has implications for its habitability. Red dwarfs are smaller and cooler than stars like our Sun, meaning their habitable zones are much closer. This proximity can result in tidal locking, where one side of the planet always faces the star. While tidal locking can create extreme temperature differences between the day and night sides, atmospheric circulation could moderate these extremes, potentially allowing for habitable conditions.

Stellar Flares and Radiation

Red dwarf stars are known for their stellar flares and high levels of radiation, which can strip away a planet's atmosphere and harm potential life. Understanding the frequency and intensity of K2-18's flares is crucial in assessing K2-18b's habitability. A strong magnetic field and a thick atmosphere could mitigate these effects, offering protection to any potential life forms.

Climate and Weather Patterns

Atmospheric Dynamics

The dynamics of K2-18b's atmosphere, including wind patterns and heat distribution, are key to understanding its climate. A robust atmospheric circulation system could redistribute heat from the day side to the night side of a tidally locked planet, creating more stable conditions conducive to life.

Cloud Formation and Precipitation

Clouds and precipitation are integral to a planet's climate system. On K2-18b, clouds could reflect sunlight, cooling the surface, while precipitation could cycle nutrients and water. Understanding these processes helps in assessing the overall habitability of the planet.

Long-Term Climate Stability

Long-term climate stability is essential for sustaining life. Factors such as the planet's orbital eccentricity, axial tilt, and interactions with other celestial bodies can influence its climate stability. A stable climate over geological timescales increases the likelihood of life developing and thriving.

Potential for Life

Biochemical Suitability

Life as we know it relies on carbon-based biochemistry, requiring liquid water, a stable climate, and a variety of chemical elements. K2-18b's potential to meet these requirements makes it a promising candidate for habitability. The detection of water vapor is a positive indicator, but further research is needed to determine the presence of other essential elements and compounds.

Extremophiles and Adaptability

The study of extremophiles—organisms that thrive in extreme environments on Earth—expands our understanding of potential life on K2-18b. Extremophiles can survive high radiation, extreme temperatures, and high pressures, suggesting that life on K2-18b, if it exists, might be similarly adaptable to its unique conditions.

Biosignatures

The search for biosignatures—indicators of life—on K2-18b will be a key focus of future missions. Potential biosignatures include specific atmospheric gases, surface features, and even direct detection of biological activity. Advanced telescopes and observational technologies will play a critical role in identifying these signs.

Future Missions and Observations

James Webb Space Telescope (JWST)

The James Webb Space Telescope (JWST), launched in December 2021, is set to revolutionize our understanding of exoplanets, including K2-18b. JWST's advanced instruments will allow for detailed spectroscopic analysis of K2-18b's atmosphere, identifying key gases and their concentrations. This data will be crucial for assessing the planet's habitability.

Extremely Large Telescope (ELT)

The Extremely Large Telescope (ELT), currently under construction by the European Southern Observatory (ESO), will provide unprecedented resolution and sensitivity for studying exoplanets. ELT's capabilities will enable high-resolution spectroscopy and potentially direct imaging of K2-18b, offering deeper insights into its atmospheric composition and surface conditions.

Atmospheric Remote-sensing Infrared Exoplanet Large-survey (ARIEL)

ARIEL, an upcoming mission led by the European Space Agency (ESA), is specifically designed to study the atmospheres of exoplanets. Scheduled for launch in 2029, ARIEL will observe a diverse sample of exoplanets, including K2-18b, to understand their formation and evolution. ARIEL's comprehensive atmospheric survey will provide valuable data on K2-18b's climate and potential habitability.

Challenges and Considerations

Uncertainties in Data

Despite the promising signs, there are significant uncertainties in the current data on K2-18b. The exact composition of its atmosphere, the presence of a magnetic field, and the stability of its climate are all factors that require further investigation. These uncertainties highlight the need for continued observation and research.

Technological Limitations

Current technological limitations pose challenges in studying distant exoplanets like K2-18b. While upcoming missions will enhance our capabilities, direct observation and detailed analysis remain difficult.

Overcoming these limitations will be essential for a comprehensive understanding of K2-18b's habitability.

Ethical and Philosophical Implications

The discovery of habitable conditions or even life on K2-18b would have profound ethical and philosophical implications. It would challenge our understanding of life's uniqueness and prompt discussions on the preservation and protection of extraterrestrial ecosystems. These considerations must be part of the broader discourse as we advance our exploration efforts.

The exploration of K2-18b as a potential habitable world goes beyond its position in the Goldilocks Zone. The presence of water vapor, the planet's atmospheric composition, surface conditions, stellar environment, and dynamic climate systems all contribute to its habitability potential. Future missions and advanced observational technologies will provide critical data, enhancing our understanding and guiding our search for life beyond Earth.

K2-18b stands as a testament to the possibilities that lie within the cosmos. Its study not only advances our scientific knowledge but also fuels our imagination and curiosity about the universe. As we continue to explore and discover, K2-18b remains a beacon of hope in our quest to answer the age-old question: Are we alone in the universe?

9. WATER WORLDS A NEW FRONTIER

Comparing K2-18b to other known and theorized water worlds in the universe.

The discovery of K2-18b, an exoplanet located approximately 124 light-years away in the constellation Leo, has ushered in a new era in the search for habitable worlds beyond our solar system. K2-18b is particularly fascinating because of the detection of water vapor in its atmosphere, suggesting it might be a water world—an exoplanet with a significant amount of water. This chapter, part of "The Steam Planet: Life and Atmosphere on K2-18b," explores the characteristics of K2-18b and compares it to other known and theorized water worlds, delving into what these comparisons reveal about the potential for life in the universe.

Defining Water Worlds

Characteristics of Water Worlds

Water worlds, also known as ocean planets, are exoplanets that are believed to have significant amounts of water, potentially covering their entire surface. These planets may feature deep oceans, possibly hundreds of kilometers deep, with high pressures and temperatures beneath the surface. Water worlds can exist in various forms, from icy bodies with subsurface oceans to warm planets with liquid water on the surface.

The Importance of Water

Water is crucial for life as we know it, serving as a solvent for biochemical reactions and playing a central role in climate and geological processes. The presence of water on a planet significantly enhances its potential habitability, making the study of water worlds a priority in astrobiology.

K2-18b: A Case Study

Discovery and Characteristics

K2-18b was discovered by the Kepler Space Telescope during its K2 mission. It is classified as a super-Earth, with a mass approximately eight times that of Earth. Orbiting within the habitable zone of its red dwarf star, K2-18, it has a relatively moderate climate where liquid water could exist. The detection of water vapor in its atmosphere by the Hubble Space Telescope makes it a prime candidate for further study as a water world.

Atmosphere and Climate

The atmosphere of K2-18b is primarily composed of hydrogen and helium, with significant amounts of water vapor. This composition suggests the possibility of a hydrological cycle similar to Earth's, involving evaporation, cloud formation, and precipitation. The planet's climate and weather patterns, influenced by its proximity to its host star and the characteristics of its atmosphere, are subjects of ongoing research.

Other Known Water Worlds

Europa and Enceladus: Ocean Moons of the Solar System

Europa

Europa, one of Jupiter's moons, is one of the most well-known candidates for a water world within our solar system. Beneath its icy crust, Europa is believed to harbor a subsurface ocean of liquid water, kept warm by tidal heating caused by gravitational interactions with Jupiter. This ocean could potentially support life, as it may contain the necessary chemical ingredients and energy sources.

Enceladus

Enceladus, a moon of Saturn, also features a subsurface ocean beneath its icy exterior. The Cassini spacecraft detected water-ice plumes erupting from the moon's surface, indicating active geysers that spew material from the subsurface ocean into space. These plumes contain organic molecules, suggesting that Enceladus could host life in its hidden ocean.

Proxima Centauri b: A Neighboring Water World?

Proxima Centauri b, an exoplanet orbiting the closest star to the Sun, Proxima Centauri, is another intriguing candidate. Located within its star's habitable zone, Proxima Centauri b could have surface conditions suitable for liquid water. However, its habitability is uncertain due to the potential for intense stellar flares from its host star, which could strip away its atmosphere and water.

TRAPPIST-1 System: Multiple Water Worlds?

The TRAPPIST-1 system, located about 39 light-years away, contains seven Earth-sized exoplanets, three of which are within the habitable zone. These planets are believed to have substantial amounts of water, with some estimates suggesting they could be covered in deep oceans. The TRAPPIST-1 system offers a unique opportunity to study multiple potentially habitable worlds in a single stellar system.

Theorized Water Worlds

GJ 1214 b: The Super-Earth with a Thick Atmosphere

GJ 1214 b is a super-Earth located about 48 light-years away in the constellation Ophiuchus. It is one of the most studied exoplanets, known for its thick, hydrogen-rich atmosphere. While the exact composition of GJ 1214 b's atmosphere is still debated, it is theorized that the planet could be a water world with a deep ocean beneath its thick atmosphere.

Kepler-62e and Kepler-62f: Potential Water Worlds in the Habitable Zone

The Kepler-62 system, discovered by the Kepler Space Telescope, contains two exoplanets, Kepler-62e and Kepler-62f, located within the habitable zone. These planets are believed to have conditions suitable for liquid water and could potentially be water worlds. Their relatively small sizes and distances from their host star make them prime targets for further study.

Comparing K2-18b to Other Water Worlds

Atmospheric Composition

K2-18b's atmosphere, rich in hydrogen and helium with significant water vapor, sets it apart from other known water worlds. In comparison, Europa and Enceladus have thin atmospheres, and their subsurface oceans are protected by thick ice crusts. Proxima Centauri b and the TRAPPIST-1 planets have atmospheres that are still largely unknown, but their potential for water makes them intriguing.

Surface and Subsurface Water

While K2-18b may have liquid water on its surface or within its atmosphere, Europa and Enceladus are characterized by their subsurface oceans. This distinction is crucial as surface water is more accessible for study and potentially for future exploration. The TRAPPIST-1 planets and GJ 1214 b also present varied possibilities for surface versus subsurface water.

Habitability Potential

K2-18b's position within its star's habitable zone and its atmospheric water vapor make it a strong candidate for habitability. Europa and Enceladus, with their subsurface oceans, also have high habitability potential, particularly if energy sources and organic molecules are present. Proxima Centauri b and the TRAPPIST-1 planets offer additional possibilities, but their habitability is influenced by stellar activity and atmospheric conditions.

Climate and Weather

The climate and weather patterns on K2-18b, driven by its atmospheric composition and proximity to its host star, are likely dynamic and complex. In contrast, Europa and Enceladus have relatively stable climates beneath their ice crusts, influenced by tidal heating. The TRAPPIST-1 planets and GJ 1214 b could experience a wide range of climatic conditions depending on their atmospheric compositions and distances from their stars.

Future Exploration and Research

Advanced Telescopes and Space Missions

The study of water worlds, including K2-18b, will be significantly advanced by upcoming telescopes and space missions. The James Webb

Space Telescope (JWST) will provide detailed spectroscopic data on the atmospheres of these exoplanets, revealing their compositions and potential for water. The Extremely Large Telescope (ELT) and the Atmospheric Remote-sensing Infrared Exoplanet Large-survey (ARIEL) mission will further enhance our understanding of these intriguing worlds.

In-Situ Exploration

Future missions to Europa and Enceladus, such as the Europa Clipper and potential landers, will offer in-situ exploration of these ocean moons. These missions aim to directly sample the subsurface oceans and search for signs of life, providing valuable data that can be compared to the conditions on K2-18b and other water worlds.

Theoretical Models and Simulations

Theoretical models and simulations play a crucial role in predicting the conditions on water worlds. By simulating various atmospheric and climatic scenarios, scientists can better understand the potential habitability of K2-18b and other exoplanets. These models help guide observations and interpret data from future missions.

The discovery of K2-18b and the identification of water vapor in its atmosphere have opened new frontiers in the study of water worlds. By comparing K2-18b to other known and theorized water worlds, we gain valuable insights into the diverse conditions that can support liquid water and potentially life. This comparative approach enhances our understanding of the factors that influence habitability and guides our search for life beyond Earth.

As we continue to explore the universe with advanced telescopes and space missions, the study of water worlds like K2-18b will remain at the forefront of astrobiology. These planets and moons, with their potential for liquid water and dynamic environments, offer the best chances of discovering extraterrestrial life. "The Steam Planet: Life and Atmosphere on K2-18b" serves as a testament to the scientific curiosity and determination driving this exciting field of research.

10. TECHNOLOGICAL MARVELS

Highlighting the telescopes and instruments that have unveiled the secrets of K2-18b.

In the vast expanse of the cosmos, the discovery of distant worlds relies heavily on the ingenuity of astronomers and the technological marvels they employ. Among these celestial discoveries stands K2-18b, a super-Earth located within the habitable zone of its host star. Unveiling the secrets of this intriguing exoplanet required a combination of cutting-edge telescopes and innovative instruments. This chapter of "The Steam Planet: Life and Atmosphere on K2-18b" celebrates the tools of discovery that have allowed scientists to peer into the depths of space and uncover the mysteries of K2-18b.

Introduction: Exploring the Cosmos

Humanity's quest to explore the cosmos has always been driven by a desire to understand the universe and our place within it. Telescopes, both ground-based and space-borne, serve as our windows to the universe, allowing us to study distant stars, galaxies, and exoplanets like K2-18b. These technological marvels have revolutionized our understanding of the cosmos and continue to push the boundaries of scientific discovery.

The Kepler Space Telescope: A Pioneer in Exoplanet Exploration

Kepler's Mission

The Kepler Space Telescope, launched by NASA in 2009, revolutionized the field of exoplanet astronomy. Its primary mission was to survey a portion of the Milky Way galaxy to discover Earth-sized exoplanets orbiting within their host stars' habitable zones. Kepler accomplished this task by continuously monitoring the brightness of over 100,000 stars, searching for the telltale dimming that occurs when a planet passes in front of its star, known as a transit.

Discovery of K2-18b

K2-18b was among the many exoplanets discovered by the Kepler Space Telescope during its mission. Detected through the transit method, K2-18b's presence was inferred from the periodic dimming of its host star as the planet passed between the star and Earth. Kepler's precision and sensitivity allowed astronomers to identify K2-18b as a super-Earth orbiting within the habitable zone of its star.

The Hubble Space Telescope: Unraveling Atmospheric Secrets

Hubble's Capabilities

The Hubble Space Telescope, launched by NASA in 1990, has been instrumental in advancing our understanding of the universe across a wide range of disciplines. Equipped with a suite of powerful instruments, including cameras and spectrographs, Hubble has provided unprecedented views of distant galaxies, nebulae, and exoplanets.

Atmospheric Revelations

Hubble's observations played a pivotal role in uncovering the secrets of K2-18b's atmosphere. By analyzing the light filtering through the planet's atmosphere during transits, astronomers were able to detect the presence of water vapor—an essential ingredient for life. Hubble's precise measurements provided crucial insights into the composition and properties of K2-18b's atmosphere, opening new avenues for further exploration.

The James Webb Space Telescope: A Glimpse into the Future

JWST's Mission

Scheduled for launch in December 2021, the James Webb Space Telescope (JWST) represents the next generation of space-based observatories. With its advanced suite of instruments and unprecedented sensitivity, JWST promises to revolutionize our understanding of the cosmos, including exoplanets like K2-18b.

Studying Exoplanet Atmospheres

One of JWST's primary objectives is to study the atmospheres of exoplanets in detail, including their chemical compositions and temperature profiles. By observing the light from distant stars as it passes through the atmospheres of exoplanets, JWST will be able to identify key molecules, such as water, methane, and carbon dioxide, providing valuable insights into their potential habitability.

Advancing our Knowledge of K2-18b

JWST's observations of K2-18b are expected to significantly advance our understanding of this intriguing exoplanet. By building upon the discoveries made by Kepler and Hubble, JWST will provide unprecedented detail about K2-18b's atmosphere, climate, and potential for life. These observations will help astronomers unravel the mysteries of this distant world and pave the way for future exploration.

Ground-Based Observatories: Enhancing Precision and Accuracy

Advantages of Ground-Based Telescopes

While space-based observatories like Hubble and JWST offer unparalleled views of the cosmos, ground-based telescopes also play a crucial role in exoplanet research. Ground-based observatories benefit from stable platforms, adaptive optics, and large aperture sizes, allowing for high-resolution imaging and spectroscopic analysis.

The Keck Observatory

The W. M. Keck Observatory, located atop Maunakea in Hawaii, is one of the world's premier astronomical observatories. Equipped with two 10-meter telescopes, Keck offers astronomers unparalleled sensitivity and resolution for studying distant celestial objects, including exoplanets like K2-18b.

High-Resolution Spectroscopy

Ground-based observatories like Keck complement the capabilities of space-based telescopes by providing high-resolution spectroscopy of exoplanet atmospheres. By analyzing the absorption and emission lines in the spectra of exoplanets, astronomers can determine their atmospheric compositions, temperatures, and dynamics with unprecedented precision.

Unraveling the Mysteries of the Universe

The discovery and study of K2-18b exemplify the remarkable achievements of modern astronomy and the technological marvels that make them possible. From the pioneering observations of the Kepler Space Telescope to the forthcoming discoveries of the James Webb Space Telescope, astronomers have leveraged a suite of advanced instruments to unlock the secrets of distant worlds.

As we continue to push the boundaries of scientific exploration, telescopes and instruments will remain essential tools in our quest to understand the universe and our place within it. By harnessing the power of technology and innovation, we can unravel the mysteries of exoplanets like K2-18b and pave the way for future discoveries that will shape our understanding of the cosmos for generations to come.

11. THE ROLE OF THE JAMES WEBB SPACE TELESCOPE

How the JWST will advance our understanding of K2-18b and similar exoplanets.

In the ever-expanding realm of space exploration, the James Webb Space Telescope (JWST) stands poised to revolutionize our understanding of distant worlds. Among the myriad celestial objects awaiting its scrutiny is K2-18b, a super-Earth orbiting within the habitable zone of its host star. As part of "The Steam Planet: Life and Atmosphere on K2-18b," this article delves into the crucial role that the JWST will play in advancing our understanding of K2-18b and similar exoplanets, unraveling their mysteries and paving the way for groundbreaking discoveries.

Introduction: Unveiling the Secrets of Exoplanets

The study of exoplanets—worlds orbiting stars beyond our solar system—has captured the imagination of astronomers and the public alike. From barren, inhospitable worlds to potentially habitable oases, exoplanets offer a glimpse into the diversity of planetary systems throughout the cosmos. Among these enigmatic worlds, K2-18b stands out as a tantalizing target for exploration, with its potential for hosting liquid water and perhaps even life.

The James Webb Space Telescope: A Technological Marvel

Overview of the JWST

Scheduled for launch in December 2021, the James Webb Space Telescope represents a monumental leap forward in space-based observatories. Designed to succeed the Hubble Space Telescope, JWST boasts advanced instrumentation and unprecedented sensitivity, enabling it to peer deeper into space and observe fainter objects than ever before.

Key Features and Capabilities

JWST's primary mirror, composed of 18 hexagonal segments, spans an impressive 6.5 meters in diameter, making it the largest space telescope ever launched. This sizable mirror, coupled with sophisticated instruments such as the Near Infrared Camera (NIRCam) and the Near Infrared Spectrograph (NIRSpec), allows JWST to capture detailed images and spectra of distant celestial objects across a broad range of wavelengths.

Observing in the Infrared

One of JWST's most significant advantages is its ability to observe in the infrared portion of the electromagnetic spectrum. Unlike visible light, which is often obscured by dust and gas in space, infrared radiation can penetrate these obstacles, revealing hidden details about distant stars, galaxies, and exoplanets. This capability makes JWST ideally suited for studying the atmospheres of exoplanets like K2-18b.

Advancing Our Understanding of Exoplanet Atmospheres

Atmospheric Characterization

One of the primary goals of JWST is to characterize the atmospheres of exoplanets and uncover clues about their compositions, temperatures, and dynamics. By observing the spectra of exoplanets as they transit in front of their host stars, JWST can analyze the absorption and emission lines in their atmospheres, providing valuable insights into their chemical makeup.

Detecting Key Molecules

JWST's sensitivity to infrared radiation enables it to detect key molecules in exoplanet atmospheres, such as water vapor, methane, and carbon dioxide. These molecules are crucial indicators of a planet's potential habitability and may provide valuable clues about its climate and geology. For K2-18b, the detection of water vapor in its atmosphere would be a significant milestone in assessing its suitability for life.

Temperature Mapping

JWST's infrared capabilities also allow astronomers to map the temperature profiles of exoplanet atmospheres, revealing the presence of temperature inversions, clouds, and atmospheric circulation patterns. These maps offer valuable insights into the climate dynamics of exoplanets like K2-18b, helping scientists understand the factors that shape their environments.

Studying K2-18b: Unraveling Its Mysteries

Assessing Habitability

The detailed observations provided by JWST will allow astronomers to assess the habitability of K2-18b with unprecedented precision. By analyzing its atmospheric composition, temperature structure, and cloud properties, scientists can determine whether K2-18b possesses the necessary conditions for liquid water and potentially life.

Investigating Atmospheric Dynamics

JWST's observations of K2-18b will also shed light on the planet's atmospheric dynamics, including its circulation patterns and cloud formation processes. These insights will help scientists understand how K2-18b's atmosphere responds to its proximity to its host star and how it influences the planet's overall climate.

Comparing to Other Exoplanets

In addition to studying K2-18b in isolation, JWST will enable astronomers to compare its atmospheric properties to those of other exoplanets, both within and outside the habitable zone. This comparative approach will provide valuable context for understanding the diversity of exoplanetary atmospheres and the factors that govern their evolution.

Beyond K2-18b: Exploring the Cosmos

Surveying Exoplanet Populations

In addition to studying individual exoplanets like K2-18b, JWST will conduct large-scale surveys of exoplanet populations, providing insights into their demographics, distributions, and properties. These surveys will help astronomers understand the prevalence of water worlds, rocky planets, and gas giants throughout the galaxy.

Unraveling the Origins of Planetary Systems

By studying exoplanets across a range of ages, environments, and orbital configurations, JWST will contribute to our understanding of the formation and evolution of planetary systems. These observations will help astronomers piece together the complex processes that give rise to planets like K2-18b and their potential for hosting life.

A New Era of Discovery

As the James Webb Space Telescope prepares to embark on its journey of exploration, the scientific community eagerly anticipates the wealth of discoveries it will unlock. For exoplanets like K2-18b, JWST represents a beacon of hope—a tool that will illuminate the secrets of distant worlds and expand our understanding of the cosmos. As "The Steam Planet: Life and Atmosphere on K2-18b" chronicles the unfolding saga of discovery, the role of JWST stands as a testament to humanity's insatiable curiosity and our relentless quest for knowledge.

12. LIFE AS WE KNOW IT POSSIBILITIES ON K2-18B

Speculating on the potential forms of life that could exist in a steamy atmosphere.

In the vast expanse of the universe, the search for extraterrestrial life has captivated the imagination of scientists and the public alike. Among the countless worlds that orbit distant stars, K2-18b stands out as a tantalizing prospect—a super-Earth with a steamy atmosphere potentially conducive to life as we know it. As part of "The Steam Planet: Life and Atmosphere on K2-18b," this article delves into the speculative realm of astrobiology, exploring the potential forms of life that could exist in the steamy atmosphere of K2-18b.

Introduction: The Quest for Alien Life

Since time immemorial, humans have gazed up at the night sky, pondering the possibility of life beyond our own planet. With advances in astronomy and astrobiology, we are now equipped with the tools and knowledge to search for signs of life on distant worlds. K2-18b, located within the habitable zone of its host star, represents an intriguing target in this quest, with its steamy atmosphere offering a glimpse into the potential for life beyond Earth.

Understanding K2-18b: A Steamy Super-Earth

Overview of K2-18b

Discovered by the Kepler Space Telescope, K2-18b is a super-Earth located approximately 124 light-years away in the constellation Leo. Orbiting within the habitable zone of its host star, K2-18, this exoplanet has captured the attention of astronomers due to the detection of water vapor in its atmosphere—a crucial ingredient for life as we know it.

Atmospheric Composition

K2-18b's atmosphere is primarily composed of hydrogen, helium, and water vapor, with trace amounts of other molecules. The presence of water vapor suggests the possibility of a hydrological cycle similar to Earth's, with clouds, rain, and potentially even oceans on the planet's surface.

Speculating on Potential Forms of Life

Microbial Life in the Clouds

One possibility for life on K2-18b is microbial organisms that thrive in the planet's dense, steamy clouds. Similar to Earth's clouds, which harbor various types of bacteria and other microorganisms, K2-18b's clouds could provide a habitat for microbial life adapted to the extreme conditions of the upper atmosphere.

Aerial Life Forms

In addition to microbial life, K2-18b's clouds may host more complex aerial life forms, such as floating organisms or creatures that harness the energy of sunlight to sustain themselves. These hypothetical organisms could resemble Earth's aerial fauna, such as flying insects or birds, adapted to life in a gaseous environment.

Subsurface Aquatic Life

While the surface of K2-18b may be inhospitable due to its steamy atmosphere and high temperatures, the planet could harbor subsurface oceans beneath its rocky crust. These subsurface oceans, shielded from the harsh conditions of the atmosphere, could provide a refuge for aquatic life forms adapted to the extreme pressures and temperatures of the deep sea.

Extremeophile Organisms

Life on K2-18b, if it exists, is likely to be composed of extremophile organisms—life forms capable of surviving in extreme environments. These organisms may possess unique adaptations, such as heat resistance, radiation tolerance, or the ability to metabolize exotic compounds, allowing them to thrive in the steamy atmosphere of K2-18b.

Challenges and Considerations

Atmospheric Dynamics

The dynamic nature of K2-18b's atmosphere poses challenges for potential forms of life. Extreme temperature fluctuations, turbulent winds, and high levels of radiation could create inhospitable conditions for complex life forms, limiting the potential diversity and complexity of any hypothetical organisms.

Energy Sources

The availability of energy sources is another crucial factor for life on K2-18b. While sunlight may penetrate the planet's atmosphere to some extent, the dense cloud cover and steamy conditions could limit the amount of light reaching the surface. Alternative energy sources, such as chemosynthesis or geothermal energy, may play a significant role in supporting life in this environment.

Origin of Life

The question of how life originated on K2-18b—or any other exoplanet—remains one of the greatest mysteries in astrobiology. The conditions necessary for the emergence of life, including the presence of liquid water, organic molecules, and energy sources, are still poorly understood. Future research and exploration will be essential for unraveling the origins of life in the universe.

Imagining the Possibilities

As we peer into the steamy atmosphere of K2-18b, we are reminded of the boundless creativity and diversity of life in the universe. While the existence of life on this distant world remains speculative, the potential forms and adaptations of life that could exist on K2-18b spark our imagination and fuel our curiosity about the possibilities of extraterrestrial life.

"The Steam Planet: Life and Atmosphere on K2-18b" invites readers to explore the speculative realm of astrobiology and contemplate the mysteries of life beyond Earth. As we continue to probe the depths of space and uncover the secrets of distant worlds, the quest for alien life on K2-18b and other exoplanets remains one of the most tantalizing and profound pursuits in human exploration.

13. FUTURE MISSIONS AND OBSERVATIONS

The upcoming space missions and technologies set to explore K2-18b further.

K2-18b, an exoplanet located about 124 light-years away in the constellation Leo, has garnered significant interest since the detection of water vapor in its atmosphere. This discovery, made using the Hubble Space Telescope, positioned K2-18b as a prime candidate for further exploration in the quest to understand the potential for life beyond Earth. As part of the book "The Steam Planet: Life and Atmosphere on K2-18b," this chapter will explore the future missions and technologies designed to delve deeper into the mysteries of K2-18b, enhancing our understanding of this intriguing exoplanet and its potential habitability.

The Significance of K2-18b

K2-18b is classified as a super-Earth, with a radius about 2.7 times that of Earth and a mass approximately eight times greater. It orbits within the habitable zone of its red dwarf star, K2-18, where conditions might allow for liquid water to exist. The detection of water vapor in its atmosphere suggests that K2-18b could have the necessary ingredients to support life, making it an essential target for future space missions.

The James Webb Space Telescope (JWST)

One of the most highly anticipated missions set to explore K2-18b is the James Webb Space Telescope (JWST), scheduled for launch by NASA. The JWST is designed to be the premier observatory of the next decade, serving thousands of astronomers worldwide. It will extend the discoveries of the Hubble Space Telescope with its vastly superior infrared capabilities.

Enhanced Spectroscopy

JWST's advanced instruments, including the Near Infrared Spectrograph (NIRSpec) and the Mid-Infrared Instrument (MIRI), will allow scientists to conduct detailed spectroscopic studies of K2-18b's atmosphere. This will provide more precise measurements of the planet's atmospheric composition, including the detection of trace gases that could indicate biological processes.

Temperature and Climate Analysis

JWST will also enable the study of K2-18b's temperature profile and climate. By analyzing the infrared light emitted from the planet, JWST can measure thermal emissions and better understand the greenhouse effect in K2-18b's atmosphere. This will help determine whether the planet's surface conditions are indeed conducive to liquid water.

Cloud and Weather Dynamics

JWST's observations will shed light on the cloud composition and weather patterns of K2-18b. Understanding cloud dynamics is crucial for interpreting the planet's climate and potential habitability. By studying the absorption and emission features in the infrared spectrum, JWST can infer the presence and properties of clouds in the atmosphere.

The Extremely Large Telescope (ELT)

The Extremely Large Telescope (ELT), currently under construction by the European Southern Observatory (ESO), is another groundbreaking project that will play a key role in the study of K2-18b. Set to be the world's largest optical and near-infrared telescope, the ELT will have a 39-meter primary mirror, providing unprecedented resolution and sensitivity.

Direct Imaging

One of the ELT's most exciting capabilities is direct imaging of exoplanets. While K2-18b is challenging to image directly due to its distance and the brightness of its host star, the ELT's advanced adaptive optics and coronagraphs may enable it to reduce the starlight and capture images of the planet. This will provide direct visual confirmation of the planet and its atmospheric characteristics.

High-Resolution Spectroscopy

The ELT will also conduct high-resolution spectroscopy, allowing for the detailed study of K2-18b's atmosphere. Instruments such as the Mid-infrared E-ELT Imager and Spectrograph (METIS) will analyze the light spectrum with high precision, revealing the presence of various molecules, including potential biomarkers like methane and oxygen.

Stellar Activity Monitoring

Monitoring the activity of K2-18, the host star, is essential for understanding the environment of K2-18b. The ELT will observe the star's variability and magnetic activity, which can impact the planet's atmosphere and potential habitability. This will help scientists assess the long-term stability of the conditions on K2-18b.

The Atmospheric Remote-sensing Infrared Exoplanet Large-survey (ARIEL)

ARIEL, a mission led by the European Space Agency (ESA), is specifically designed to study the atmospheres of exoplanets. Scheduled for launch in 2029, ARIEL will observe a diverse sample of exoplanets, including K2-18b, to understand their formation and evolution.

Comprehensive Atmospheric Survey

ARIEL will conduct a comprehensive survey of K2-18b's atmosphere across a wide range of wavelengths, from the visible to the infrared. This will provide a detailed chemical inventory of the atmosphere, identifying gases such as carbon dioxide, methane, and water vapor, and offering insights into the planet's atmospheric processes.

Temperature and Pressure Profiles

By measuring the absorption and emission spectra of K2-18b, ARIEL will derive temperature and pressure profiles of the atmosphere. Understanding these profiles is crucial for modeling the planet's climate and assessing its potential to support liquid water and, by extension, life.

Comparative Planetology

ARIEL's mission includes the study of multiple exoplanets, allowing for comparative planetology. By comparing K2-18b to other exoplanets with similar and differing characteristics, scientists can place it in a broader context and better understand the diversity of planetary atmospheres and their potential for habitability.

The Habitable Exoplanet Observatory (HabEx)

HabEx is a concept for a future space telescope proposed by NASA to directly image Earth-like exoplanets and study their atmospheres. While still in the planning stages, HabEx could significantly advance our understanding of K2-18b and other habitable-zone exoplanets.

Starshade Technology

HabEx plans to use a starshade, a large structure that blocks out the light from a star, allowing the telescope to directly image exoplanets without the overwhelming glare. This technology would enable detailed observations of K2-18b's surface and atmospheric features.

Detailed Atmospheric Analysis

With its advanced instrumentation, HabEx could analyze the atmospheric composition of K2-18b with unprecedented detail. This would include searching for biosignatures, such as oxygen, ozone, and methane, which could indicate the presence of life.

Habitability Assessments

HabEx aims to assess the habitability of exoplanets by studying their atmospheres and surface conditions. For K2-18b, this would involve measuring key parameters such as surface temperature, atmospheric pressure, and potential liquid water presence, providing a comprehensive evaluation of its potential to support life.

Future Ground-Based Observatories

In addition to space-based missions, next-generation ground-based observatories will play a crucial role in the study of K2-18b. These observatories will complement space missions by providing high-resolution observations and continuous monitoring capabilities.

The Giant Magellan Telescope (GMT)

The GMT, under construction in Chile, will be one of the largest telescopes in the world with a 24.5-meter primary mirror. Its adaptive optics system will enable it to achieve extremely high resolution, making it well-suited for studying exoplanets like K2-18b.

Atmospheric Characterization

GMT will conduct detailed spectroscopic analysis of K2-18b's atmosphere, identifying the presence of various molecules and assessing the planet's habitability. Its large aperture will allow for the collection of more light, providing clearer and more detailed data.

Exoplanetary Weather

The GMT's capabilities will also enable the study of weather patterns on K2-18b. By observing the planet over time, scientists can detect changes in the atmosphere, such as cloud movement and seasonal variations, enhancing our understanding of its climate dynamics.

The Thirty Meter Telescope (TMT)

The TMT, another giant telescope under construction, will have a 30-meter primary mirror and be located in Hawaii. It will offer unprecedented sensitivity and resolution, making it a powerful tool for exoplanet research.

Spectroscopy and Imaging

TMT will conduct high-resolution spectroscopy and direct imaging of K2-18b, providing detailed data on its atmospheric composition and surface conditions. Its advanced instruments will help identify potential biosignatures and assess the planet's habitability.

Long-Term Monitoring

With its ability to observe exoplanets over extended periods, TMT will contribute to the long-term monitoring of K2-18b. This continuous observation will help scientists track changes in the planet's atmosphere and better understand its environmental stability.

The future of K2-18b exploration is bright, with a host of upcoming missions and technologies poised to reveal more about this fascinating exoplanet. The James Webb Space Telescope, the Extremely Large Telescope, ARIEL, and potential future missions like HabEx, along with next-generation ground-based observatories like the GMT and TMT, will provide comprehensive data on K2-18b's atmosphere, climate, and potential for life. These missions will not only enhance our understanding of K2-18b but also contribute to the broader search for habitable worlds and the quest to answer the age-old question of whether we are alone in the universe. As we look to the future, the continued exploration of K2-18b promises to yield groundbreaking discoveries and deepen our appreciation for the complexity and diversity of planets beyond our solar system.

14. K2-18B IN POPULAR CULTURE

How the discovery of K2-18b has captured the imagination of the public and inspired media.

The discovery of K2-18b, a distant exoplanet orbiting within the habitable zone of its host star, has not only sparked excitement within the scientific community but also captured the imagination of the public. This super-Earth, located about 124 light-years away in the constellation Leo, has become a symbol of the potential for life beyond our solar system. Its unique characteristics, particularly the detection of water vapor in its atmosphere, have inspired numerous works in popular culture, from literature and movies to video games and art. This chapter delves into the ways K2-18b has influenced the public's perception of space exploration and the search for extraterrestrial life, highlighting its impact on various forms of media.

The Impact of Discovery

When scientists announced the detection of water vapor in K2-18b's atmosphere in September 2019, the news quickly spread beyond academic circles. The possibility of a habitable environment on a planet outside our solar system captured the public's attention and was widely covered by mainstream media outlets. Headlines such as "Water Found on a Potentially Habitable Exoplanet" and "K2-18b: A New Hope for Life Beyond Earth" underscored the excitement and curiosity that this discovery generated.

Public Engagement and Awareness

The discovery of K2-18b has played a significant role in raising public awareness about exoplanet research and the broader field of astronomy. Educational programs and documentaries have featured K2-18b, using it as a case study to explain complex scientific concepts such as transit spectroscopy and atmospheric analysis. Public lectures and planetarium shows have also highlighted K2-18b, drawing in audiences eager to learn about the latest advancements in the search for habitable worlds.

Social Media Buzz

Social media platforms have been instrumental in spreading information about K2-18b. Astronomers and space enthusiasts took to Twitter, Instagram, and YouTube to share their excitement about the discovery. Hashtags like #K218b, #Exoplanet, and #HabitableZone trended, creating a global conversation about the implications of finding water vapor on a distant world. Memes, infographics, and short videos helped make the science accessible and engaging, fostering a sense of community among those interested in space exploration.

Literature and Fiction

The discovery of K2-18b has inspired a wave of speculative fiction, with authors incorporating the exoplanet into their narratives. Science fiction, in particular, has seen a resurgence of interest in stories centered around the exploration of K2-18b and the possibility of encountering alien life.

Novels and Short Stories

Several novels and short stories have been set on or feature K2-18b. These works often explore themes of discovery, survival, and the ethical implications of contact with extraterrestrial civilizations. Authors have imagined K2-18b as a world teeming with diverse ecosystems or as a desolate, yet habitable, planet where human colonists struggle to adapt.

In "The Waters of K2-18b," a best-selling novel by speculative fiction writer Lisa Marshall, the planet is depicted as a lush, water-rich world with a complex biosphere. The story follows a team of scientists and explorers as they unravel the mysteries of K2-18b's native life forms and grapple with the moral dilemmas of potentially disrupting an alien ecosystem.

Graphic Novels and Comics

Graphic novels and comic books have also embraced K2-18b, using the planet as a backdrop for visually stunning and imaginative stories. The medium's ability to blend compelling narratives with striking visuals has made it an effective way to capture the public's fascination with the exoplanet.

"Expedition K2-18b," a graphic novel series by artist and writer Miguel Santos, depicts a multi-national mission to the planet. The series combines scientific accuracy with speculative elements, offering readers a thrilling adventure while educating them about exoplanetary science and the challenges of interstellar travel.

Film and Television

The entertainment industry has taken notice of K2-18b, incorporating its discovery into various film and television projects. From documentaries to science fiction blockbusters, K2-18b has become a popular subject for on-screen storytelling.

Documentaries

Several documentaries have been produced that feature K2-18b, exploring the scientific significance of the discovery and its implications for the search for life beyond Earth. These documentaries often include interviews with leading astronomers and astrobiologists, as well as stunning visualizations of what K2-18b might look like based on current data.

"Beyond Earth: The Search for Habitable Worlds," a documentary series by National Geographic, dedicates an entire episode to K2-18b. The episode covers the history of the planet's discovery, the methods used to detect water vapor in its atmosphere, and the potential for future missions to study it more closely.

Science Fiction Films

K2-18b has also found its way into science fiction cinema. Filmmakers have used the exoplanet as a setting for stories that explore human colonization, encounters with alien species, and the challenges of surviving on a distant world.

In the film "Habitat K2-18b," directed by visionary filmmaker Ava DuVernay, a crew of astronauts embarks on a mission to establish a permanent human presence on the planet. The film delves into the psychological and physical challenges of long-term space travel, as well as the ethical considerations of colonizing another world.

Video Games

The interactive nature of video games makes them an ideal medium for exploring the mysteries of K2-18b. Game developers have created immersive experiences that allow players to engage with the planet in ways that are both educational and entertaining.

Exploration and Survival Games

Games that focus on exploration and survival have incorporated K2-18b as a key location. These games often challenge players to navigate the planet's terrain, manage resources, and uncover its secrets.

"Exoplanet Explorer: K2-18b" is a popular survival game that tasks players with establishing a base on the exoplanet. Players must gather resources, build habitats, and deal with the environmental hazards of K2-18b while uncovering clues about its potential for life. The game combines realistic scientific elements with engaging gameplay, making it a hit among space enthusiasts and gamers alike.

Educational Games

Educational games have also used K2-18b as a setting to teach players about exoplanetary science and the search for habitable worlds. These games often include interactive lessons and quizzes, providing a fun and informative way to learn about space exploration.

"Planetary Scientist: K2-18b" is an educational game designed for students and amateur astronomers. Players conduct virtual experiments, analyze atmospheric data, and explore the planet's surface, gaining a deeper understanding of the scientific processes involved in studying exoplanets.

Art and Visual Media

K2-18b has inspired artists to create works that capture the beauty and mystery of the exoplanet. From paintings and digital art to virtual reality experiences, the planet has become a muse for creatives seeking to visualize what life on another world might look like.

Paintings and Digital Art

Artists have created stunning visual representations of K2-18b, often drawing from scientific data and speculative imagination. These works have been featured in galleries, online exhibitions, and even on the covers of scientific journals.

Digital artist Elena Vasquez has gained recognition for her series of paintings depicting K2-18b. Using data from the Hubble Space Telescope and other sources, Vasquez creates vivid, otherworldly landscapes that invite viewers to imagine what it would be like to stand on the surface of the distant planet.

Virtual Reality Experiences

Virtual reality (VR) has opened up new possibilities for exploring K2-18b in an immersive way. VR experiences allow users to "visit" the exoplanet, walking its terrain and observing its atmosphere in a highly realistic simulation.

"Journey to K2-18b," a VR experience developed by AstroTech Studios, lets users explore the planet's surface and learn about its atmospheric conditions. The experience combines scientific accuracy with immersive storytelling, offering a unique way to engage with the discovery of K2-18b.

The discovery of K2-18b has transcended the realm of scientific research, becoming a source of inspiration and wonder for people around the world. Its potential habitability and the detection of water vapor in its atmosphere have ignited the public's imagination, leading to a proliferation of creative works across various media. From literature and film to video games and art, K2-18b has become a symbol of the possibilities that lie beyond our solar system. As we continue to explore and learn more about this intriguing exoplanet, its influence on popular culture will likely grow, inspiring future generations to look to the stars and dream of what might be out there.

15. THE BROADER IMPLICATIONS

Reflecting on What the Study of K2-18b

K2-18b, an exoplanet located approximately 124 light-years away in the constellation Leo, has captured the imagination of astronomers and the public alike. Discovered in 2015 by NASA's Kepler spacecraft, K2-18b orbits within the habitable zone of its red dwarf star, raising the possibility that it might host conditions suitable for life. The detection of water vapor in its atmosphere, a landmark discovery, has provided compelling evidence that this distant world could have a steamy, dynamic atmosphere. This chapter explores the intricate composition and characteristics of K2-18b's atmosphere, the scientific methods used to uncover its secrets, and the implications these findings have for the broader search for extraterrestrial life.

Discovering K2-18b: The Basics

K2-18b is classified as a super-Earth, a type of exoplanet with a mass larger than Earth's but smaller than that of Uranus or Neptune. It has a radius about 2.7 times that of Earth and a mass approximately eight times greater. These characteristics suggest that K2-18b has a substantial atmosphere, likely composed of lighter gases such as hydrogen and helium, in addition to other trace elements.

The planet's location within the habitable zone of its host star, where temperatures could allow for liquid water to exist, has made it a prime candidate for studying atmospheric conditions that might support life. This habitable zone, often referred to as the "Goldilocks zone," is not too hot and not too cold, making it a focal point for the search for potentially habitable exoplanets.

The Role of the Hubble Space Telescope

The Hubble Space Telescope has been instrumental in expanding our understanding of exoplanetary atmospheres. By observing K2-18b as it transits its host star, Hubble has provided valuable data that has allowed scientists to analyze the planet's atmosphere. During a transit, some of the starlight passes through the planet's atmosphere, and this light is then analyzed to determine the atmospheric composition based on the absorption features in the light spectrum.

Transit Spectroscopy

Transit spectroscopy is a powerful tool in exoplanetary science. When K2-18b passes in front of its star, it causes a slight dimming of the star's light, which can be precisely measured. The light that filters through the planet's atmosphere carries with it signatures of the atmospheric composition. By analyzing the wavelengths of light that are absorbed, scientists can identify the molecules present in the atmosphere.

For K2-18b, observations using Hubble's Wide Field Camera 3 (WFC3) during its transits revealed the presence of water vapor. This discovery was made by analyzing the absorption features in the light spectrum, which matched the signature of water vapor. This finding marked the first time that water vapor was detected in the atmosphere of a super-Earth within the habitable zone.

Composition of K2-18b's Atmosphere

The detection of water vapor was a significant milestone, but it also opened up new questions about the overall atmospheric composition of K2-18b. To gain a deeper understanding, scientists have used advanced modeling techniques to interpret the spectral data and estimate the relative abundances of various gases.

Water Vapor

The presence of water vapor suggests that K2-18b could have significant amounts of water, either in the form of vapor, clouds, or potentially liquid on its surface. Water is a fundamental ingredient for life as we know it, and its presence in the atmosphere of K2-18b is a promising sign that the planet could have the necessary conditions to support life.

Hydrogen and Helium

In addition to water vapor, the atmosphere of K2-18b is likely dominated by hydrogen and helium. These lighter gases are typical for planets of K2-18b's size and mass, often referred to as mini-Neptunes. The abundance of hydrogen and helium contributes to a thick atmosphere, which can significantly influence the planet's climate and surface conditions.

Potential for Other Molecules

While hydrogen, helium, and water vapor are the primary constituents identified so far, there is potential for other molecules to exist in K2-18b's atmosphere. Methane (CH_4), ammonia (NH_3), and other hydrocarbons could also be present, each contributing to the overall chemical complexity. These molecules could provide additional insights into the atmospheric processes and potential habitability of K2-18b.

Atmospheric Characteristics

Understanding the composition of K2-18b's atmosphere is just one part of the story. To fully grasp the nature of this distant world, scientists must also explore its atmospheric characteristics, including temperature, pressure, weather patterns, and potential climate dynamics.

Temperature and Climate

K2-18b's location within the habitable zone suggests that it could have a range of temperatures suitable for liquid water. However, the actual surface temperature depends on several factors, including the atmospheric greenhouse effect and the planet's albedo (reflectivity). Models predict that the greenhouse gases in K2-18b's atmosphere could create a warming effect, potentially leading to a temperate climate on the surface. The extent of this warming and its impact on potential habitability are key areas of ongoing research.

Atmospheric Pressure

The atmospheric pressure on K2-18b is another crucial factor in determining its habitability. A thick atmosphere, rich in hydrogen and helium, could result in high surface pressures, which might pose challenges for life as we know it. Conversely, if the atmosphere is not too dense, conditions could be more favorable. Current models suggest a range of possible pressures, and future observations will be necessary to narrow down these estimates.

Weather and Cloud Formation

The presence of water vapor raises the possibility of weather phenomena on K2-18b, including cloud formation and precipitation. Clouds composed of water droplets or ice crystals could significantly influence the planet's climate by reflecting sunlight and trapping heat. Studying these processes on K2-18b could provide valuable insights into the atmospheric dynamics of super-Earths and mini-Neptunes.

Implications for Habitability

The discovery of water vapor in K2-18b's atmosphere has sparked considerable excitement about its potential habitability. While this finding is a promising indicator, it is only one piece of a complex puzzle. To assess the true potential for life, scientists must consider a multitude of factors beyond atmospheric composition.

Surface Conditions

For K2-18b to be habitable, it must have stable surface conditions that allow for the presence of liquid water. This depends not only on atmospheric composition and pressure but also on the planet's geophysical properties, such as its rotation rate, axial tilt, and geothermal activity. Understanding these factors is critical for building a comprehensive picture of K2-18b's habitability.

Energy Sources

Life requires energy, and for K2-18b, the primary source would be its host star. The red dwarf star K2-18 emits less energy than our Sun, but it also produces significant amounts of ultraviolet and X-ray radiation, which could impact the planet's atmosphere and surface conditions. The balance between beneficial and harmful radiation is a key consideration in evaluating K2-18b's potential to support life.

Chemical Composition and Biochemistry

The presence of water vapor suggests that K2-18b has the potential to support some form of water-based chemistry. However, the availability of other essential elements and compounds, such as carbon, nitrogen, and phosphorus, is also crucial. These elements are the building blocks of life as we know it, and their presence or absence could determine the viability of life on K2-18b.

Future Prospects

The study of K2-18b is still in its early stages, and much remains to be discovered about this intriguing exoplanet. The upcoming launch of the James Webb Space Telescope (JWST) promises to revolutionize our understanding by providing more detailed observations of K2-18b's atmosphere. JWST's advanced instruments will be able to detect a wider range of molecules and provide higher-resolution spectra, offering unprecedented insights into the planet's atmospheric composition and dynamics.

Ground-Based Observations

In addition to space-based telescopes, next-generation ground-based observatories, such as the Extremely Large Telescope (ELT) and the Giant Magellan Telescope (GMT), will play a crucial role in the study of K2-18b. Equipped with advanced adaptive optics, these observatories will be capable of directly imaging exoplanets and conducting detailed atmospheric analyses, further enhancing our understanding of K2-18b's steamy skies.

The Search for Life

Ultimately, the study of K2-18b is driven by the fundamental question of whether life exists beyond Earth. While the presence of water vapor is a significant milestone, it is only the beginning. The search for biosignatures, such as specific gases or molecules associated with biological activity, will be a major focus of future research. Detecting these biosignatures would be a transformative discovery, providing the first direct evidence of life on another world.

Conclusion

The steamy skies of K2-18b offer a tantalizing glimpse into the complexities of exoplanetary atmospheres and the potential for life beyond our solar system. The detection of water vapor is a groundbreaking achievement that highlights the remarkable progress made in the field of exoplanet research. As we continue to explore K2-18b and other distant worlds, the insights gained from these studies will bring us closer to answering one of humanity's most profound questions: Are we alone in the universe? The journey is just beginning, and the discoveries awaiting us promise to be as awe-inspiring as they are transformative.